Introduction to Permaculture

Introduction to Permaculture

Rachel Santiago

SYRAWOOD
PUBLISHING HOUSE

New York

Published by Syrawood Publishing House,
750 Third Avenue, 9th Floor,
New York, NY 10017, USA
www.syrawoodpublishinghouse.com

Introduction to Permaculture
Rachel Santiago

International Standard Book Number: 978-1-64740-073-6 (Hardback)

Cataloging-in-Publication Data

Introduction to permaculture / Rachel Santiago.
    p. cm.
Includes bibliographical references and index.
ISBN 978-1-64740-073-6
1. Permaculture. 2. Alternative agriculture. 3. Agricultural ecology. I. Santiago, Rachel.
S494.5.P47 I58 2022
631.58--dc23

TABLE OF CONTENTS

Permissions

Index

This book aims to help a broader range of students by exploring a wide variety of significant topics related to this discipline. It will help students in achieving a higher level of understanding of the subject and excel in their respective fields. This book would not have been possible without the unwavered support of my senior professors who took out the time to provide me feedback and help me with the process. I would also like to thank my family for their patience and support.

Permaculture refers to the conscious maintenance and design of agriculturally productive ecosystems that have the stability, diversity and resilience of natural ecosystems. There are various branches of permaculture such as environmental design, ecological engineering, regenerative design and ecological design. It also focuses on integrated water resource management that helps in the development of sustainable architecture. Some of the common practices of permaculture are hügelkultur, natural building, rainwater harvesting, sheet mulching, grazing, Keyline design and fruit tree management. The major principles of permaculture are care for the Earth, and the people. This book aims to shed light on some of the unexplored aspects of this field. The topics herein on this field are of utmost significance and bound to provide incredible insights to readers. Through this book, we attempt to further enlighten the readers about the new concepts in this field.

A brief overview of the book contents is provided below:

Chapter – What is Permaculture?

A system of principles that draw from patterns and resilient features of nature is referred to as permaculture. These principles are extensively applicable in the field of regenerative agriculture, rewilding and community resilience. This is an introductory chapter which will briefly introduce all the significant aspects of permaculture.

Chapter – Concepts of Permaculture

Permaculture comprises of a wide range of concepts which are important in various domains. Assisted natural regeneration, polyculture, conservation agriculture, holistic management, conservation grazing, regenerative agriculture, rotational grazing, etc. are some of the concepts. This chapter discusses in detail these concepts related to permaculture.

Chapter – Common Practices in Permaculture

There is a diverse range of practices that are followed in permaculture. Natural building, agroforestry, hügelkultur, fruit tree pruning and management, intensive rotational grazing, etc. are some practices that fall in its domain. The topics elaborated in this chapter will help in gaining a better perspective about different practices in permaculture.

Chapter – Components of Permaculture

Permaculture consists of numerous components such as folkewall, composting toilet, leaf mold, mulch, Three Sisters, treebog, spent mushroom compost, groundcover, etc. This chapter closely examines these key components of permaculture to provide an extensive understanding of the subject.

Chapter – Techniques used in Permaculture

Companion planting, forest gardening, grassed waterway, holzer permaculture, mycoforestry, intercropping, keyline design, raised-bed gardening, sheet mulching, vegan organic gardening, companion planting, etc. are the example of various techniques used in permaculture. The aim of this chapter is to explore these techniques of permaculture.

Chapter – Other Fields Related to Permaculture

Permaculture is a vast subject that finds its applications in various other fields. Some of them include organic farming, integrated farming, community gardening, sustainable development, rainwater harvesting, etc. This chapter has been carefully written to provide an easy understanding of the varied fields related to permaculture.

Chapter – Environmental Design

A design that aims to improve and enhance natural, social, cultural and physical aspects of the surroundings of a specific area is termed as environmental design. Agroecology, landscape design and planning and layers of forest such as canopy, understory, shrub layer, etc. are among its components. All the diverse components related to environmental design have been carefully analyzed in this chapter.

Chapter – Desert Greening

Man-made restoration of deserts for the purposes of farming, forestry or reclamation of natural water systems to support life, is defined as desert greening. It deals with methods such as, farmer-managed natural regeneration, afforestation, reforestation, etc. The topics elaborated in this chapter will help in gaining a better perspective about desert greening.

Rachel Santiago

What is Permaculture?

- Principles of Permaculture
- Zones in Permaculture
- Benefits of Permaculture
- Applications of Permaculture

A system of principles that draw from patterns and resilient features of nature is referred to as permaculture. These principles are extensively applicable in the field of regenerative agriculture, rewilding and community resilience. This is an introductory chapter which will briefly introduce all the significant aspects of permaculture.

Permaculture integrates land, resources, people and the environment through mutually beneficial synergies – imitating the no waste, closed loop systems seen in diverse natural systems. Permaculture studies and applies holistic solutions that are applicable in rural and urban contexts at any scale. It is a multidisciplinary toolbox including agriculture, water harvesting and hydrology, energy, natural building, forestry, waste management, animal systems, aquaculture, appropriate technology, economics and community development.

Permaculture (the word, coined by Bill Mollison, is a portmanteau of permanent agriculture and permanent culture) is the conscious design and maintenance of agriculturally productive ecosystems which have the diversity, stability, and resilience of natural ecosystems. It is the harmonious integration of landscape and people — providing their food, energy, shelter, and other material and non-material needs in a sustainable way. Without permanent agriculture there is no possibility of a stable social order.

Permaculture design is a system of assembling conceptual, material, and strategic components in a pattern which functions to benefit life in all its forms.

The philosophy behind permaculture is one of working with, rather than against, nature; of protracted and thoughtful observation rather than protracted and thoughtless action; of looking at systems in all their functions, rather than asking only one yield of them; and allowing systems to demonstrate their own evolutions.

As the basis of permaculture is beneficial design, it can be added to all other ethical training and skills, and has the potential of taking a place in all human endeavors. In the broad landscape, however, permaculture concentrates on already settled areas and agricultural lands. Almost all of these need drastic rehabilitation and re-thinking.

One certain result of using our skills to integrate food supply and settlement, to catch water from our roof areas, and to place nearby a zone of fuel forest which receives wastes and supplies energy, will be to free most of the area of the globe for the rehabilitation of natural systems. These need never be looked upon as "of use to people", except in the very broad sense of global health.

The real difference between a cultivated (designed) ecosystem, and a natural system is that the great majority of species (and biomass) in the cultivated ecology is intended for the use of humans or their livestock. We are only a small part of the total primeval or natural species assembly, and only a small part of its yields are directly available to us. But in our own gardens, almost every plant is selected to provide or support some direct yield for people. Household design relates principally to the needs of people; it is thus human-centered (anthropocentric).

This is a valid aim for settlement design, but we also need a nature-centered ethic for wilderness conservation. We cannot, however, do much for nature if we do not govern our greed, and if we do not supply our needs from our existing settlements. If we can achieve this aim, we can withdraw from much of the agricultural landscape, and allow natural systems to flourish.

Recycling of nutrients and energy in nature is a function of many species. In our gardens, it is our own responsibility to return wastes (via compost or mulch) to the soil and plants. We actively create soil in our gardens, whereas in nature many other species carry out that function. Around our homes, we can catch water for garden use, but we rely on natural forested landscapes to provide the condenser leaves and clouds to keep rivers running with clean water, to maintain the global atmosphere, and to lock up our gaseous pollutants. Thus, even anthropocentric people would be well-advised to pay close attention to, and to assist in, conservation of existing forests and to assist in, the conservation of all existing species and allow them a place to live.

We have abused the land and laid waste to systems we never need have disturbed had we attended to our home gardens and settlements. If we need to state a set of ethics on natural systems, then let it be thus:

- Implacable and uncompromising opposition to further disturbance of any remaining natural forests, where most species are still in balance;

- Vigorous rehabilitation of degraded and damaged natural systems to stable states;

- Establishment of plant systems for our own use on the least amount of land we can use for our existence;

- Establishment of plant and animal refuges for rare or threatened species.

Permaculture as a design system deals primarily with the third statement above, but all people who act responsibly in fact subscribe to the first and second statements. We believe we should use all the species we need or can find to use in our own settlement designs, providing they are not locally rampant and invasive.

Vegan Permaculture

Veganism is defined as "a philosophy and way of living which seeks to exclude, as far as is possible and practicable, all forms of, and cruelty to, animals for food, clothing or any other purpose; and by extension, promotes the development and use of animal-free alternatives for the benefit of humans, animals and the environment".

Of course, we cannot exist on Earth without causing some form of harm, whether it is from walking on tiny insects, or driving in our cars, or building any type of structure, however, we can do these things more mindfully and reduce our impact as much as possible.

Vegan permaculture embodies the ethics of permaculture at a deeper, more authentic level because the ethics of veganism becomes the driving force and foundation behind the permaculture ethics. This means that vegan permaculture rejects the use of animals or any animal derived product through any type of exploitation within its design systems or thinking. It is more than organic gardening too as it encompasses the whole system, integrating everything and segregating nothing.

A balanced ecosystem involves plants and wildlife. They work in harmony with each other as the natural world is meant to. Humans can learn so much from the natural world, especially in terms of our individual footprint on the planet. When we take into consideration our individual impact simply by being alive, we become more aware of the importance of our choices regarding everything we do. These choices include all our consumables, from food, clothing, accessories, furniture, etc., and how are they produced. For example, spending your hard earned money on purchasing food grown under modern agriculture practises, even organic, still involves the use and abuse of exploited animals such as their manure or blood and bone. Other choices also cover our mode of transport, the locations of where we live and work, our building designs, and the communities we live in. The list is ongoing.

Humans can live more in harmony with each other and the planet by becoming more aware and responsible for themselves and their choices. Each individual is powerful; we can make significant differences to our modern world through our everyday choices.

Vegan Permaculture for Human Health: People Care

Human health is deteriorating around the globe. There is a huge amount of scientific evidence that our diets play a massive role in preventing, stopping and reversing some of our most debilitating illnesses and causes of death. When we're young we don't notice the long term impact of eating animal derived products but as we age we realise we're not as 'bullet-proof' as we once thought we were. Most people accept this as a part of life but then they resort to either heavily medicating, which has its own side effects, and invasive surgery or treatments. All of these practises are extremely dangerous and age us even more.

Eating a wholefoods, plant-based diet gives us the best chance at optimal health, especially if we accompany that with exercise and a balanced work-life ratio.

Another element to consider is our mental and spiritual health. If we think about the energetic and vibrational difference between a place of slaughter to a place of fun and happiness, we can 'feel' a definite distinction. Non-vegans take into their bodies and spirit all the negative energy and stress that farmed and hunted animals experience, especially leading up to and including their slaughter. Not only that, humans know on a deep level that we are all connected, and that murder for pleasure (which includes convenience, taste, tradition and habit) is wrong. The guilt, confusion and dislike we ultimately feel for ourselves by partaking in these atrocious acts towards our fellow Earthlings is subconscious, and ultimately has a negative impact on our state of mind and spiritual vibration. Consider too the vegan produce grown in modern agriculture practises, it also bears a burden.

Once people make the reconnection with their own compassion, usually lost or suppressed during childhood, they can then live in a more authentic way that gives them a greater sense of meaning to their lives. Vegans too can be unaware of where their food comes from and how it is produced, so to grow as much as we can ourselves or source it with full transparency provides the cleanest, healthiest and most peaceful food on the planet. We all can empower ourselves and be our most authentic selves by becoming more aware and more responsible.

Vegan Permaculture for the Animals: Other Earthling Care

Animals suffer immensely at the hands of humans. We have enslaved certain species of animals and bred them for us to use and exploit, along with the many wild species that we have also hunted/captured. However animals are not ours to use for food, fashion, experimentation, entertainment, labour or bred to be our companions. All animals are our fellow Earthlings and have the right to live their lives free from harm or exploitation by humans for humans or other animals. Animals deserve this freedom regardless of whether we consider them cute or intelligent, or whether we find them useful, or any other 'humanised' opinion about them that we may have.

We humans must change our thinking and attitudes so that we can change our actions to truly do as little harm as we possibly can.

Vegan Permaculture for the Environment: Earth C Permaculture tries to create resilient living systems that are inspired by processes, structures it with enough resources (nutrients) to restart development again.

There is a huge amount of science and research proving that animal agriculture is destructive to the environment. Whether it is from deforestation for grazing farmed animals or growing their food, to fresh water consumption, to waste and pollution, to species extinction and habitat destruction, there is no doubt that animal agriculture has an adverse effect upon planet Earth and all species.

While traditional permaculture teaches how to farm plants and animals on our own properties or locally within communities as a more sustainable, ethical and environmental way of living, vegan permaculture totally excludes the use of farmed animals. Instead it teaches how to design and utilise only plants and wildlife within any food production system, creating only natural ecosystems where wildlife are the only animal contributors. This allows humans to be in harmony with nature and wildlife, causing the least amount of negative environmental impact. Vegan permaculture can be incorporated in both urban and rural locations, in food production, human settlement design and planning, community living, and regenerating the natural world.

Principles of Permaculture

Permaculture tries to create resilient living systems that are inspired by processes, structures, and patterns observed in nature. Design principles have emerged that are used as a framework for the design of complex agroecosystems. Some permaculture designers have developed their own sets of principles, depending on the focus of their work. The most commonly used set of permaculture principles was developed by co-originator, David Holmgren. These twelve principles are presented as the result of an 'in-depth analysis of the natural environment and pre-industrial and sustainable societies, the application of ecosystem theory, and design thinking'. They claim to provide a framework for the design of sustainable land use and a society within ecological boundaries.

The principles are short statements that point the way when dealing with complex systems and give a variety of options for action. The first six principles use a bottom-up approach, while the final six principles can be seen from a top-down designer's perspective. Also, because of this, some overlaps between the principles occur. Trying to produce no waste and applying self-regulation will lead to integration rather than segregation of elements, or observing and interacting empowers to be able to creatively respond to change. In the design process, it is important not to focus on one or few principles, but to use the set as a whole and create a balance within the system.

Although Ferguson and Lovell have recognised the isolation of permaculture from science, we hypothesize that there is strong scientific evidence for the individual principles, underlining their applicability in the redesign of agricultural systems towards sustainability.

Table: Summary of the twelve permaculture principles proposed by permaculture co-originator, David Holmgren, with corresponding approach (bottom-up or top-down), relation (design process, management, agroecosystem structure), and examples with scientific evidence presented in this issue.

Principle	Approach	Relation	Examples with Evidence
I. Observe and Interact	bottom-up	Design process, management	Adaptive management
II. Catch and Store Energy	bottom-up	Agroecosystem structure	Organic mulch application Rainwater harvesting measures Woody elements in agriculture.
III. Obtain a Yield	bottom-up	Design process, management	Emergy evaluation Ecosystem services concept.
IV. Apply Self-Regulation and Accept Feedback	bottom-up	Agroecosystem structure	Enhancement of regulating ecosystem services Natural habitats in agricultural landscapes Wildflower strips.
V. Use and Value Renewable Resources and Services	bottom-up	Agroecosystem structure	Legumes and animal manure as nutrient source Mycorrhizal fungi.
VI. Produce no Waste	bottom-up	Agroecosystem structure	Animal manure Human excreta Waste products as animal feed.
VII. Design from Patterns to Details	top-down	Agroecosystem structure, Design process	Natural ecosystem mimicry Use of grazing animals in cold and dry climates Structurally complex agroforests in tropical climates.
VIII. Integrate Rather than Segregate	top-down	Agroecosystem structure	Integration of livestock in corn cropping Cereals and canola used for forage and grain harvest Integration of fish in rice cropping Polyculture (crops).

IX. Use Small and Slow Solutions	top-down	Agroecosystem structure	Inverse productivity-size relationship Agroforestry systems.
X. Use and Value Diversity	top-down	Agroecosystem structure	Plant species diversity Pollinator diversity Habitat diversity Diversified farming systems.
XI. Use Edges and Value the Marginal	top-down	Agroecosystem structure	High field border density Field margins Edges with forests.
XII. Creatively Use and Respond to Change	top-down	Design process, management	Decision-making under uncertainty Increase ecological resilience Directed natural succession.

Permaculture Principle I: Observe and Interact

This principle stands for the method of alternating observation and interaction with a certain system to generate knowledge and experience about it. The scientific management approach related to this principle is called adaptive management, which is a systematic approach for improving resource management by learning from management outcomes. Therefore, multiple management options to reach specific management goals are implemented. The monitoring of system responses to management options gives decision guidance to adjust management practice. Adaptive management was, for instance, successfully used to investigate and improve the effectiveness of agro-environmental schemes in protecting the corn bunting, Emberiza calandra.

Simulation results show that an adaptive management approach yields the best trade-off between agricultural production and environmental services in the case of severe drought in vineyards. However, this approach, with its emphasis on feedback-learning to face the unpredictability and uncertainty that is intrinsic to all ecosystems, is not new. In some traditional management systems, the direction of resource management is guided by the use of local ecological knowledge to interpret and respond to feedback from the environment. While there are still barriers, like maintaining long-term monitoring, to establish adaptive management, the great potential to improve our understanding of important ecological processes necessary for effectively managing biological systems is already visible. Investigations of grazing systems also indicate that the lack of adaptive management in scientific experiments explains why those trials were not able to reproduce the positive effects reported by experienced practitioners.

Permaculture Principle II: Catch and Store Energy

Different sources of energy are covered by this principle: E.g., solar energy, water, wind, living biomass, and waste. According to this principle, energy shall be held within the system as long as possible. This is necessary to be able to use it as long and effectively as possible and to maintain their functions, such as buffering extreme events. The most important storages of future value are fertile soil with high humus content, perennial agroecosystems (especially trees), and water storages, such as groundwater and water bodies.

One method to catch and store energy in the form of water, nutrients, and organic matter, while protecting the existing storage of fertile soil, is to apply organic mulch. The application of mulch greatly increases soil water storage efficiency, as well as the water use efficiency of crops, therefore, also increasing crop yields. The application of mulch also leads to higher organic matter content in the soil and therefore enhances microbial biomass, soil microbial functional diversity, and nitrogen cycling. Higher contents of soil organic matter lead to higher and more stable yields in agriculture. One explanation is the higher capacity of organic matter rich soils to buffer drought stress. Soil fertility is directly linked to soil organic matter. Experiments show that without maintaining natural nutrient cycling via litter decomposition, and without supplementary fertilization, agriculture is only economical for 65 years on temperate prairie, for six years in tropical semi-arid thorn forest, and for no more than three years in Amazonian rainforest. At the same time, the application of mulch showed to be highly effective in preventing soil erosion achieved by reducing runoff and increasing infiltration. The capture and storage of rainwater in the soil can be enhanced through rainwater harvesting (RWH) measures. These include linear contour structures, like bunds or grass strips, terracing, semi-circular bunds, and pitting. This mainly helps to overcome drought events, leading to increased food security and income for farmers. However, RWH measures also enhance ecosystem services, such as groundwater recharge, nutrient cycling, and biodiversity.

The incorporation of woody elements, such as trees, shrubs, and hedges, into agriculture also represents an application of this principle, amongst others, through the storage of carbon. Different options of land stewardship have a high potential for climate change mitigation, out of which reforestation has the greatest overall potential, and the incorporation of trees in croplands has one of the highest potentials for agriculture and grasslands. Apart from a highly needed climate change mitigation potential, those measures also provide benefits, such as habitats for biodiversity, enhanced soil and air quality, and improved water cycling.

Permaculture Principle III: Obtain a Yield

The (farming) systems designed and managed with permaculture have to obtain a sufficient yield, and to supply humans with food, energy, and resources. However, this principle also aims at the efficiency of production, as our "yield" is low if we have to put in a lot of effort, energy, and resources to obtain it. Apart from that, this principle also calls for a more holistic understanding of yield, not only an economic one, but also ecologic and social yields.

Emergy analysis is a value-free environmental accounting method based on a holistic systems concept, which is suitable to measure the yield of agro-ecosystems in this sense of efficiency. Emergy is defined by Howard Odum as the available energy of one kind that has already been used to make a product or provide a service. It is usually measured in solar emergy joules (sej), and allows the calculation of various indices. The emergy yield ratio (EYR) provides information on how much emergy output is generated by the system per input of the economy, while the renewability (REN) gives the share of emergy of an output that is provided by renewable natural resources or services. Recent research shows that our modern food production systems are highly inefficient in terms of resource and energy consumption. Corn production in the USA had an EYR of only 1.07 with an REN of 5%, while the EYR of conventional pig production was even lower at 1.04 and an REN of 26%. Numerous methods used in permaculture are derived from indigenous people, such as the traditional Lacandon Maya agroecosystems in Mexico. These systems cycle through three stages of production, starting with field crops, progressing to shrubs and then to the trees, before returning to field crops. Therefore, they direct natural succession and are able to yield resources from a polyculture with as many as 60 plant species and without inputs of seeds, fertilizer, or pesticides. For six of those systems analysed, the EYR ranged from 4.5 to 50.7 with an REN ranging from 0.72 to 0.97, indicating a high level of sustainability. However, land productivity in terms of the calories of these systems is much lower compared to modern corn production. As this principle also calls for obtaining a yield to feed the people, a combination of efficiency and sustainability, as well as land productivity, should be aspired to. However, this combination is probably the most difficult point in agroecological systems, at least when compared to modern, industrial agricultural production. At first sight, this looks like an approval of the 'land-sharing' approach. However, permaculture is a site-specific and context-based design system. Therefore, we would conclude that it depends on the context of the farm and region whether a 'land-sharing', a 'land-sparing', or a combination of both approaches is most favourable to reach the goal of ensuring the resilience of the whole system, while producing enough food.

The call for a more holistic understanding of yield associated with this principle is comparable with the concept of ecosystem services. Scientists try to use this concept to value ecological as well as cultural services to advance the appreciation of non-monetary services provided by nature. Hereby, a holistic understanding of yields from ecosystems is also demanded.

This principle is especially crucial as it calls for a sufficient yield of agricultural products while maintaining a high efficiency in terms of resource and energy consumption as well as ecological and social 'yields'. To further investigate this principle, research has to be carried out, including the evaluation of land productivity of permaculture or similar systems. This is crucial to evaluate whether such systems, besides improving ecological functioning, have the potential to feed the growing world population.

Permaculture Principle IV: Apply Self-Regulation and Accept Feedback

The goal of permaculture is to create systems as self-sustaining and self-regulating as possible. Positive feedback accelerates growth and energy accumulation within the farming systems. This is best used in the early phase. Negative feedback, the more important one, protects the system from instability or scarcity through miss- or over-usage. Additionally, each element within a land use system should be as self-reliant as possible to increase the resilience against disturbances.

The enhancement of regulating ecosystem services, such as natural pest control, pollination, nutrient cycling, and soil and water quality regulation, are the most common applications of this principle. Strengthening of stabilizing feedbacks in ecological systems, such as those regulating ecosystem services, helps to maintain a favoured and resilient regime of the ecosystem and increases robustness against external stress, e.g., climate change.

In the case of insect pollination, this ecosystem service is jointly responsible for a stable (low variability) yield of dependent crops. Increasing the proximity to and the sharing of natural habitats might be one way to apply self-regulation, as this increases the temporal and spatial stability (high predictability and low variation during the day and among plants, respectively) of the pollination service.

The reintroduction of flower rich habitats into the agro-ecosystem is another measure to apply self-regulation through the enhancement of the stabilizing ecosystem service of natural pest control. Perennial, species-rich wildflower strips were able to enhance natural pest control and thereby to increase yields from adjacent wheat fields by 10%. A reduced need for pesticide use leads to a lower impact on biodiversity and thereby again increases ecosystem stability through biodiversity-related ecosystem services, such as pollination and pest control. At the same time, the dependency of the farm on agrochemicals decreases, making the farm itself more self-reliant.

Permaculture principle V: Use and Value Renewable Resources and Services

The use of renewable resources and services is necessary to stop the exploitation of non-renewable resources, which, in the long run, undermines the functionality of the whole system. Plants might be used as an energy source, building material, and soil improvers, while examples for animals are herding dogs, animals for soil cultivation, and draught animals. This principle also covers the use of wild resources (fish, game, wood), which should be used sustainably to maintain the renewability of these resources. Overall, this principle focuses on maximizing the use and functioning of ecosystem services.

One well studied example for this principle is the use of nitrogen fixing plants (legumes) or animal manure instead of mineral nitrogen fertilizer. Firstly, mineral nitrogen fertilizer contributes 40–68% to farm energy demand and thereby greatly increases the net global warming contribution of farming systems. Alternatively, the energy demand of legume nitrogen fixation is provided by solar radiation and animal manure is available as a waste product. At the same time, animal manure and legume-based systems show higher yield resilience to drought stress as well as increased soil carbon stocks. Animal manure is also proposed to be a renewable resource to stop micronutrient depletion of soils, which is already interlinked with malnutrition of the population in some regions of the world. It has to be kept in mind that there are trade-offs concerning these alternatives to mineral fertilizer. Legumes reduce land use efficiency when they are only used to replace fertilizer and are not harvested as a crop. In the long run, animal manure is only renewable if animal production, including feed production, is based only on renewable resources.

Other renewable service providers linked to this issue are mycorrhizal fungi. Mycorrhizal fungi are more abundant with organic fertilization, while mineral nitrogen fertilization decreases the diversity of mycorrhizal fungi. Mycorhizzas increase the water and nutrient uptake of plants and thereby enhance plant growth and yield, especially under drought conditions.

There are only a few scientific results dealing with working animals as renewable resources. Most of them are investigating draught animals in countries of the global South. Results indicate that primary energy consumption is lower when using cattle for ploughing compared to tractors. However, we could not find sufficient scientific evidence for the comparison of animal work with machinery in agriculture. Important issues to be investigated according to this comparison are energy efficiency and resource consumption, labor productivity, and environmental impacts, such as soil erosion and greenhouse gas emissions.

Permaculture Principle VI: Produce No Waste

This principle aims at mimicking the natural pattern of exchange and cycling of matter and energy. In natural living systems, no waste occurs as every output of an element (a species) is used by another element. This is why waste could also be seen as an output, which is not used by the system.

According to this, all waste should be seen as a resource that should be used to be as effective as possible.

The most important example for this principle from modern agriculture is possibly animal manure. Through the separation of plant and animal production in industrial agriculture, animal manure became a waste and a problem. This is due to huge animal production systems concentrating in some regions, while feeds, depending on fertilizer input, are produced elsewhere. Through land application, the high amount of animal manure produced in some regions leads to environmental problems, like eutrophication of ground and fresh water, heavy metal accumulation in top soils, and the emission of ammonia, greenhouse gases, and noxious odours. However, recent studies show that in other regions and in lower concentrated land, the application of animal manure has huge benefits. It is a valuable resource to enhance plant nutrient availability (including micronutrients), water holding capacity, soil structure, organic matter content, and carbon storage. Even if animal manure is applied on agricultural land at reasonable rates, storage and transportation can still cause environmental problems, such as ammonia and greenhouse gas emissions. By designing smaller and integrated agricultural systems, animal manure can lose its waste character while maintaining a high quality and fertile soil.

Even more important might be the high amount of waste of human excreta. Human excreta is a valuable resource of nutrients needed in agricultural production. The application of human excreta as fertilizer is still common in parts of the world, e.g., in Vietnam. Urine is especially valuable as it contains 50–90% of the nutrients of human excreta. It also has a high hygienic quality as it only contains few enteric microorganisms. To improve the hygienic quality of human faeces, a common practice is composting human excreta, which can strongly decrease pathogenous bacteria and parasites. However, land applications of human excreta is still a critical topic as there is still insufficient evidence for the fate of therapeutic agents.

Many other examples of using waste products in agriculture have been documented. Feeding a 10% share of dried grape pomace increases the growth performance and health indicators of lambs. Vegetable and fruit waste occurring in high amounts with industrial production could also be used as animal feed.

Permaculture Principle VII: Design from Patterns to Details

Natural ecosystems should be used as patterns for sustainable land use as natural ecosystems evolved over a long period of time to function under certain environmental conditions. Additionally, landscape patterns, such as geomorphology, catchments, and methods, like zoning, and sectors should be used in permaculture design for effective site planning.

In scientific literature, this principle is known as "natural ecosystem mimicry". The main patterns/ models that are usable for agricultural ecosystems are grasslands, such as savanna or prairie, dry forests, and tropical rainforests. Large areas on earth are naturally too cold or too dry for agriculture. These are areas where natural grasslands occur as the climate is also not suitable for trees. The natural pattern found here is grasslands crossed by large herds of grazing animals. The strategy here is to use the grazers on natural vegetation and to harvest them for meat (e.g., cattle, sheep, goat) or their metabolites (e.g., milk). Some authors suggest the application of the natural pattern of densely packed and continuously moving herds through multi-paddock, rotational, cell, or mob grazing to prevent desertification through grazing. Areas that are dry, but able to facilitate some trees, normally inhabit savannah like systems. In addition to grazers (or in some cases already crops), trees are included to maintain ecosystem functioning, such as a hydrological balance similar to dry forests. In temperate regions, forests might be used as models for agroecosystems by combining perennial woody crops, such as nut and fruit trees, with different kinds of animals, such as cattle, sheep, poultry, and pigs.There are also areas that are too wet to be suitable for agriculture, namely, the humid tropical lowlands. The trophic complexity of local biota (including pests) and nutrient leaching limit agricultural suitability. The strategy here is to increase usable productivity while maintaining the natural structure by building diverse and structurally complex agroforests containing mostly perennial species, as it has been done successfully by local people for centuries.

For other planning strategies, such as zoning and sectoring, no scientific evidence could be found. Further research has to be carried out to investigate whether those planning strategies lead to higher labor productivity through improved farm logistics, as well as higher performance and resilience of agricultural elements through site-specific positioning.

Permaculture Principle VIII: Integrate rather than Segregate

Biological interactions, especially mutual ones, should be used to increase the productivity and stability of the agroecosystem and to generate synergy effects. Integration of elements enables making use of the multifunctionality of elements, like chickens for pest control when integrated into an orchard system. Integration also allows sustaining important functions of a system through multiple elements, like chickens and fruit trees both covering the function of food production. This leads to higher stability of the agroecosystem through integrated pest control and higher economic resilience as the yield is distributed to two sources.

These benefits of re-integrating elements in agriculture, especially crops and livestock, have also been promoted in the scientific literature. This integration of crops and livestock is proposed to help overcome the dichotomy between the increase in agricultural production and the negative environmental impacts. This is achieved through better regulation of biogeochemical cycles, an increase in habitat diversity and trophic networks, and a greater resilience of the system against socio-economic or climate change induced risks and hazards. Case studies from France and Brazil show that increasing the interactions between subsystems decreased dependence on external inputs and increased the efficiency of the farm, leading to a good economic, as well as environmental, performance and an increased resilience against market shocks. Studies from the USA show that integrating livestock in corn cropping systems via cool season pasture significantly increases soil quality indicators, such as organic matter and nutrient content. In Australia, the dual-purpose use of cereals and canola for forage during the vegetative stage while still harvesting for grain afterwards is practiced. This provides risk management benefits, improves soil properties, and is able to increase both the livestock and crop productivity of farms by 25–75% with little increase in inputs. Additionally, findings from Asia show that the integration of fish into rice cropping systems increases crop yields through improved weed and pest regulation, increased nutrient availability, and improved water flows, while additionally yielding fish without additional feed or fertilizer. Furthermore, a meta-analysis identified a strong potential of within-field crop diversification (polyculture) for win-win relationships between the yield of a focal crop species and the biocontrol of crop pests.

A recent study on 35 self-identified permaculture farms in the United States shows, that most of them rely on mixed annual and perennial cropping, the integration of perennial and animal food crops or even an the integration of production and services such as education.

Permaculture Principle IX: Use Small and Slow Solutions

This principle is derived from a fundamental pattern found in living organisms: Cellular design. Functions are covered on the smallest possible level, while larger-scale functions are provided through replication and diversification. This principle includes the assumption that small-scale systems are potentially more intense and productive (such as marked gardening or gardening for self-sufficiency), while slow growing systems are potentially more stable and effective (such as tree-based systems).

Small farms (1–2 ha) cultivate 12% and even smaller family farms (less than 1 ha) cultivate 72% of the word's agricultural land, and therefore secure nutrition for the biggest share of the world's population. In the scientific literature, the relationship between farm size and land productivity (output per area) has been widely investigated. An inverse productivity-size relationship, stating that smaller farms are more productive per area, has been observed in Africa, Asia, Europe, and Latin America. Smaller farms, and therefore field sizes, also lead to a higher amount of field edges, inducing beneficial effects, which will be discussed with principle 11.

Modern arable farming systems undermine ecosystem functioning through the adverse effects of intensive industrial production, such as soil erosion, climate change, and loss of biodiversity. Agroforestry systems are slower developing compared to modern arable farming, taking some years to reach full productivity and profitability. However, through the maintenance of ecosystem services, such as erosion control, climate change mitigation, biodiversity, and soil fertility, they maintain ecosystem functioning. At the same time, agroforestry systems are proposed to be more resilient to climate change. In the long run, agroforestry systems have the potential to be even more productive, when compared to exclusively agricultural systems.

The application of animal manure or legumes as fertilizer is another example of a slower solution, compared to the fast availability of nutrients from synthetic or mineral fertilizer. Long term studies show that it takes some years until manure or legume fertilized systems (in this case corn) reach a comparable productivity to systems fertilized with mineral fertilizers. However, in the end, manure and legume fertilized systems were both more resistant to draughts, maintaining their yields in drought years. As mentioned above the application of manure also maintains and enhances soil quality and fertility.

As this principle is also aimed at farm setup and development, it is also necessary to investigate from an economical perspective whether small and slow developing farms are more economically stable. Recent results of a case study in France show that it is possible for one person to earn a living from agriculture with relatively low input (e.g., no motorization) on 0.1 ha when using permaculture.

Permaculture Principle X: Use and Value Diversity

This principle is based on the assumption that diversity is one of the foundations of adaptability and the stability of ecosystems. This is why, also in agroecosystems, the habitat and structural diversity should be maintained, as well as the age, species, variety, and genetic diversity.

Many ecosystem services maintaining the functioning of our agroecosystems are related to biodiversity. A meta-analysis shows that increasing biodiversity, in many cases of plant species, has positive effects on productivity in terms of producer and consumer abundance, on erosion control through increased plant root biomass, on nutrient cycling through increased mycorhizza abundance and decomposer activity, and on ecosystem stability through increased consumption and invasion resistance. It has also been shown that increasing pollinator diversity has significant positive effects on the yields of various pollination dependent crops. Habitat diversity, in terms of landscape complexity, has positive effects on ecological pest control. As an example, increasing the habitat and flowering plant diversity through artificial wildflower strips can increase yields by 10% in nearby wheat fields through enhanced ecological pest control.

The increasing awareness of the importance of this principle – to use and value diversity—can also be seen in the development of diversified farming systems. Diversified farming systems use practices developed via traditional and agroecological scientific knowledge to intentionally include functional biodiversity at multiple spatial and temporal scales. Several studies show that these attempts in increasing agrobiodiversity and therefore ecosystem services, such as soil quality, carbon sequestration, water-holding capacity in surface soils, pollination, pest control, energy-use efficiency, and resistance and resilience to climate change, are successful.

Permaculture Principle XI: Use Edges and Value the Marginal

Edges are potentially more diverse and productive, as resources and functions of both adjacent ecosystems are present. As in agroforestry systems, these edge zones can be increased on purpose to take advantage of this effect. Edge zones can also be planned as an appropriate separation of elements, such as woody strips in between meadows. This principle is also aimed at valuing margins for their often invisible advantages and functions instead of trying to minimize them.

Recent scientific results show that increasing farmland configurational heterogeneity (higher field border density) increases the pollination ecosystem service through higher wild bee abundance and an improved seed set of test plants, probably through enhanced connectivity. Investigations at the former Iron Curtain in Germany, where the East switched to large-scale farming while the West maintained small-scale agriculture with >70% longer field edges, show similar results. Here, higher biodiversity was found in the region with small scale agriculture, while the species richness and abundance were also higher in field edges compared to field interiors, indicating a link between biodiversity increase and field edge density.

Beyond field edge densities, field margins, often seen as unproductive areas, are also of great importance in maintaining ecosystem services. Margins have a range of associated fauna, some of which may be pest species, while many are beneficial either as crop pollinators or as pest predators, and therefore contribute to the sustainability of production by enhancing beneficial species within crops and reducing pesticide use.

Edges with other ecosystems may even have a stronger effect on ecosystems services supporting the agroecosystem. Pollination of coffee in terms of fruit set increased in the transition zone to forest ecosystems due to higher functional pollinator diversity, while the quantity and quality of strawberries was higher near pond edges through a higher abundance of pollinators. Increasing edges with other, especially natural, ecosystems leads, in most cases, to a fragmentation of those habitats. Habitat fragmentation is often associated with habitat loss, which has large negative effects on biodiversity. However, an investigation of 118 studies on habitat fragmentation, independent of habitat amount, showed that 76% of significant biodiversity responses to habitat fragmentation were positive. Negative effects of habitat fragmentation per se are likely due to habitat size becoming too small to sustain a local population (e.g., mammalian predators) or to negative edge effects (e.g., increased predation of forest birds at edges).

As edges appear to have positive effects, it should be mentioned that edges also have the potential to produce negative effects on agricultural production. In transition zones from forests to agricultural areas, changes in the microclimate and matter cycling occur, some of which are not favorable for crop production, such as shade and resource competition.

Permaculture Principle XII: Creatively use and Respond to Change

Natural ecosystems are stable and resilient despite constant change and the influence of disturbances. The potential for evolutionary change is essential for the dynamic stability of ecosystems. That is why such systems should not be considered as being in a fixed state, but as an evolutionary process. The implications for agroecosystem design are to include flexibility to create resilience and to deliberately use natural change, such as succession.

Our earth's ecosystems are complex, which means that their responses to human use are generally not linear, predictable, or controllable. Another property of complex systems is the existence of momentum, leading to a temporal dynamic of the system. Coupled with the high replication time of ecological experiments, this limits applied ecological research, leading to a permanent existence of uncertainties associated with ecological systems. Therefore, it is essential when dealing with ecological systems to apply decision theory's principles of decision-making under uncertainty, which will not be worked out in detail here. To be able to creatively use and respond to change, systems need to be monitored and assessed. Actions should also be favored that are reversible and robust to uncertainties, and that increase the resilience of the (socio-)ecological system. Ecological resilience can be defined as the magnitude of disturbance that an ecosystem can withstand without changing self-organized processes and structures. In general, ecological resilience is based on two pillars: The diversity of habitats, species, and genes and reservoirs, such as fertile soil, water, or biomass.

In the case of natural succession, one example of how to use the dynamics and changes in natural ecosystems through successive planting and the facilitation of usable annuals, herbaceous perennials, shrubs, and trees has already been given. In Mexico, indigenous people use and direct natural succession to create a highly efficient land use system. Another example on a much smaller temporal scale is rotational or cell grazing, where only a short, but intense, pulse of disturbance is used to set the grassland system back to an earlier stage of succession, leaving it with enough resources (nutrients) to restart development again.

Zones in Permaculture

The design principle of Zones and Sectors is concerned with efficient energy planning, that is, planning the placement of elements in the design, such as trees and plants, animals, structures and buildings, to make to most efficient use of energy.

Efficient energy planning can be broken down to the following three categories:

- Zone Planning.
- Sector Planning.
- Slope.

Zone Planning

Zone planning is a system where the location of an element in a design is determined by:

- How often we need to use the element?
- How often we need to service the element?

This is a basic logical principle, whereby the things you use most often, and the things you have to pay the most attention to, are placed closest to the house in the design.

Consequently, the things that are used the least often, or that require little or no attention, are placed furthest away in the design, and things that fall somewhere in between are placed accordingly.

By situating the most often used or serviced elements in a design closest to the home, it makes it easier to access them. This means less energy is expended to access them, making for a more energy efficient design.

As a practical example, a kitchen garden containing the most often used vegetables and herbs would ideally be located in close proximity to the kitchen itself, so when the need for herbs and vegetables arises, it's only a quick step outside the back door of the house to get the required cooking ingredients. It would be highly inefficient, and extremely wasteful of energy if you had to walk across your whole property, to some remote back corner, to get what you need to prepare a meal, for the following reasons. Firstly, you'd be less likely to go there, and secondly, you would have difficulty maintaining a kitchen garden that is harder to access – as you can't keep an eye on it, and are less inclined to maintain it if it takes a whole lot of effort and energy to do so simply because it is so far out of the way.

Zones are abstract conceptual boundaries around the home which help us to work with distance to plan efficient energy use.

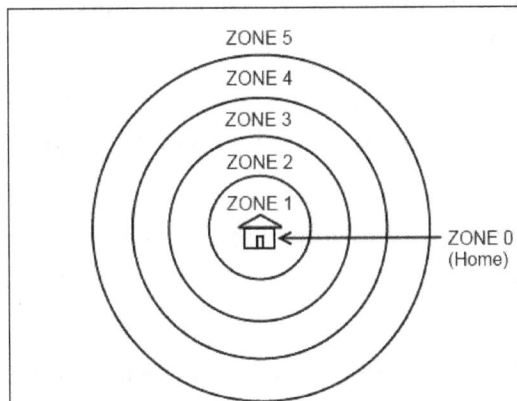

The areas around a house are divided up into zones numbered 0 to 5, based their accessibility and frequency of use in relation to the location of the house. The lowest number denotes the most frequently accessed areas, while the highest number indicates the areas least accessed.

Here is a conceptual zone diagram, illustrating the various zones around the house.

Zones are often misunderstood in Permaculture design. Stress the following points before we get into further details:

- Zones are not hard boundaries, they are not necessarily delineated by fences or other hard structures.

- Zones can blend into each other at their boundaries, this is most often the case in real life designs.

- Zones are not circular, they can be any shape, defined by how accessible the areas around the house are.

By defining the different zones around the home, we can then create some guidelines of what we can put in each zone in our designs:

Zone 0

- This is the home itself, the centre of activity.

- Unless you are creating a Permaculture design for a bare block of land, and have to decide where to locate the house, and the design house itself, then Zone 0 is not normally a concern for most designers.

- Where there is a pre-existing house on the land, that normally will be your Zone 0 and the beginning of your zone mapping. If you do have to locate and design the house, It goes without saying that the home design should be energy efficient, and provides an environment where the occupants can live and work in a sustainable and harmonious way.

Zone 1

- This is the most intensively used zone, and the most managed and controlled.

- Zone 1 is the area nearest to the house, and also includes the most frequently accessed areas, such as alongside often used paths.

- Keep in mind that this zone is defined by access, so if there is an area near the house that you don't visit, or is hard to get to, even if it sits next to the house itself, then it is not included in Zone 1.

- If you leave your property daily to go go work for example, then the path from the street to your house and the immediate areas alongside it will be included in Zone 1, as you visit these areas twice daily.

- The diagram below shows how the area around the house, coloured green, is Zone 1, as is the path from the street. The area to the the side of the shed is not an area that is often accessed, so it is not included in the Zone 1 region, even though it is very close to the house.

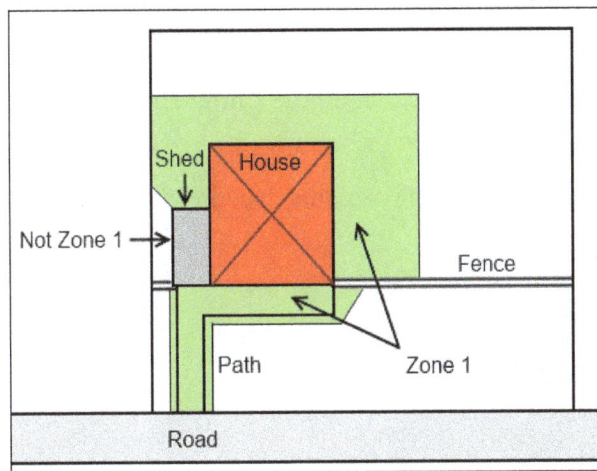

Elements that are located in this zone include all the things that you need to access most often, or that need the most frequent attention, such as:

- A kitchen garden to provide vegetables and salad greens which have a short growing season (time from planting to harvest) and herbs for teas and culinary use.

- Small trees which provide often used fruit, such as lemons.

- Worm farms for processing kitchen waste.

- Greenhouses, cold frames and propagation areas.

- Workshops or sheds.

- Rainwater tanks, water bores and wells.

- Fuel for heating or fire, such as wood or gas.

- Small animal pens/cages for rabbits or guinea pigs.

Zone 1 plantings usually employ complete mulching, using a system such as sheet mulching, and are fully irrigated with irrigation systems such as drip systems, which sit below the mulch on the garden beds. Zone 1 is an intensive system, it is a human ecology, that does not exist in Nature, and would fall apart without human attention.

- For many urban properties, where the backyard is quite small, then the whole property will be Zone 0 and Zone 1 only.

- Most urban properties can only be divided up into Zones 0, 1 and 2, and the designs for these properties will not need to include any of the other zones.

- When starting to map zones on your site design, start with Zone 1 first, and when constructing the site, build and complete Zone 1 first, then work outwards to complete the other zones.

Zone 2

This zone is also used quite intensively, but a bit less than Zone 1, and accommodates some of the larger and slightly less frequently used elements, that still need fairly frequent attention.

Elements that are located in this zone include all the things that you need reasonably often, or that need the fairly frequent attention, such as:

- Perennials and vegetables which have a long growing season (time from planting to harvest).

- Fruit trees/orchards.

- Compost bins.

- Bee hives.

- Ponds.

- Chicken/poultry enclosures.

- Enclosures for larger animals that need to be regularly monitored, maintained and attended to.

Zone 2 plantings can employ complete mulching using a system such as sheet mulching, but if the area is too large and this is impractical. then spot mulching around the trees may be employed, and tree guards can be used to protect trees while they get established. These plantings are fully irrigated using irrigation systems such as drip systems.

Zone 3

The zone is basically farmland, where the main crops are grown (for personal use and to sell), where orchards of larger trees are located, and where livestock is kept and grazed. Once these areas are established, they only require minimal maintenance and care.

Elements that are located in this zone include all the things that require infrequent attention only, such as:

- Orchards of larger trees.

- Main farming crops.

- Pastures and rearing areas for large livestock such as cows and sheep.

- Semi-managed bird flocks.

- Large trees for animal forage – oak trees and nut trees.

- Dams for water storage and drinking water for animals.

Zone 3 plantings employ green mulching, which is an under-planting of ground cover plants which serve as a 'living mulch' for the trees. These plantings are unpruned, and not all plants have irrigation to water them.

Zone 4

This zone is a part wild/part managed, and its main use is for collecting wild foods, timber production, as a source of animal forage, and pasture for grazing animals.

The trees in this zone are managed by allowing animals to browse to control new growth, or by thinning (removing) seedlings to select the variety of trees that will be allowed to grow.

Zone 5

This zone is an unmanaged wild natural ecosystem, such as bushland, forest or similar natural area, free of human intervention, interference or control.

Zone 5 is a wilderness conservation area, and space that provides us with the opportunity to step down from our role of controlling Nature, to one where we can just witness Nature in its pure form, where we can simply observe the cycles of Nature and learn from what we see. It the place where we can meditate and reconnect with Nature, and come to understand our place in the world.

The wilderness area does not have to be restricted to the outer perimeters of a property in a design. Zone 5 can extend as a wedge all the way from the outer perimeter right up to the house, to create a wildlife corridor as part of a design that brings natural ecosystem close the the home.

In urban areas, Zone 5 can be a nearby creek, or a neglected area of unused vacant land.

Practical Zone Diagrams

Now that we have discussed some of the guidelines of what we place in each zone, it is appropriate to now revisit our zone diagram, but with a more practical focus.

The reason zones are rarely circular is because ground is rarely flat, and even apparently flat ground will have a measurable gradient. Furthermore, areas of land can be irregularly shaped, so real world zone diagrams can appear very different from our previous conceptual zone diagram.

Here is an example of a zone diagram which is closer to a real-life example, where each zone is shown in a different colour for illustrative purposes.

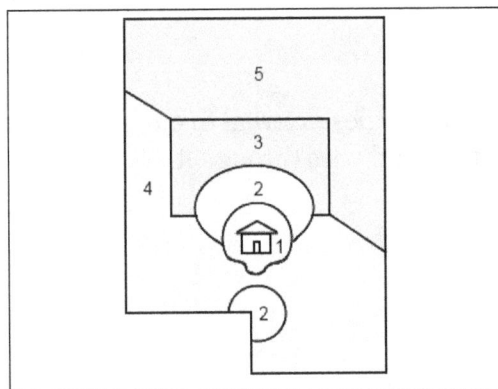

Here, we can see that the zones can be irregularly shaped, they can overlap rather than form concentric circles and a particular zone can appear more than once.

This should illustrate the flexibility we have in mapping zones in zone diagrams, and how far from the circular conceptual diagram real-life examples can be.

Zone Sizes

One question that often arises for designers is "how big should the zones be"?

The size of a zone is driven by two factors:

- The distances that are practical to cover on a human scale,

- The amount of space required to yield produce to support a given number of people.

With these factors in mind, here are some practical design guidelines for the ideal amount of area allocated to each zone.

Zone 1: Is ideally around 1000 sq. m (1/4 acre) in size for a family of four, this size is manageable as an intensive food production system. (All vegetables required can be grown in an area of 50sq. m per person.)

Zone 2: Is ideally 4000 sq. m (1 acre) in size for a family.

Zone 3: Can range from 4 to 20 acres for a family.

Zone 4: Can be any size.

Zone 5: Is a wilderness and is used for hunting and gathering.

In conclusion, zones are concerned with the flow and use of energy inside our system, optimising it by the use of distance, and the strategic placement of elements, according to their frequency of use and the attention they require.

Zone planning though, does not account for all the systems of energy interacting with the site we are designing. A site does not exist in isolation, it exists as part of a larger environment, where external energies, the elements of nature, which come from outside our system, also act on it.

To plan for these energy systems, we use a system of energy planning known as Sector Planning.

Sector Planning

Sector planning is concerned with energies external to our site, the elements and forces of Nature, that come from outside our system, and pass through it.

These energies include:

- Hot summer winds.

- Cold winter winds.

- Winter and summer sun angles.

- Salty or damaging winds.

- Water flow and flood prone areas.

- Unwanted views.

- FIre danger areas.

Since, these wild energies come into our system from outside, we can strategically place elements in our design to manage or take advantage of these incoming energies.

By placing plants, trees or structures in the appropriate areas, we can:

- Block the incoming energy.

- Channel the incoming energy for our intended use.

- Open the area to allow the incoming energy in.

Blocking Incoming Energy

Where external incoming energy is detrimental to our system, we can block its flow, preventing disruption to our system.

Wind is an element which often requires steps to manage it in most designs. Hot summer winds, cold winter winds, salty seaside breezes, and damaging dusty winds all need to be restricted in a design through the use of windbreaks. Windbreaks can be constructed using specifically resilient plants and trees, or by building protective structures.

Identifying where the summer sun and winter sun shines is important for managing the harsh midday and afternoon summer sun (north and west sun in southern hemisphere, south and west sun in northern hemisphere). Deciduous trees can be planted around the house to block the sun in summer, keeping the house cool. In winter, when the leaves fall, the low winter sun can warm the house naturally. Man-made structures can also be built around the house which take advantage of the sun's low winter angle and high summer angle to provide summer shade and winter sun.

Where fire dangers exist, the areas most prone to incoming fire are identified, and firebreaks are placed in this area in our designs. We place elements here that do not burn such as roads, cleared areas, stony ground, concreted areas, stone walls, ponds, marshes and waterways. These areas are planted with fire-resistant tree species and vegetation to create a shelter belt. Trees suitable for this purpose are typically European deciduous trees, such as deciduous fruit and shade trees. A selection of suitable trees includes deciduous fruit trees in general, oaks, elms, willows, poplars, aspens, cottonwoods, figs, carob, mulberries and mirror bush.

Another application of 'blocking incoming energy' is the screening of unwanted views. Trees, plants and structures can be erected to provide additional privacy, and block out unwanted views, while providing a more aesthetically pleasing alternative.

Channelling Incoming Energy for our Use

Free energy coming into our site from outside can also be utilised for our benefit.

Water flowing into our site, either from directly above as rain, from run-off coming from adjacent properties, or collecting in an area (such as a flood prone area) can be redirected into lakes, dams, ponds, irrigation channels, swales and other water management systems.

Wetter areas can be used specifically to grow very 'thirsty' plants and trees, which will help manage the excess water, or they can be converted to wetlands or bodies of water, such as ponds, lakes and dams for water storage.

Water can be captured at an elevated point on the site, and being elevated, it is a store of what they call 'potential energy' in physics. The water can then flow under gravity to perform work, such as irrigation or water supply.

Water flowing across a stream or river can be used to drive a hydroelectric generator to provide electricity, or can have some of the flow diverted for irrigation purposes.

Wind can be captured to drive wind turbines or windmills, providing a source of free energy to the site which we can utilise for our purposes.

Sunlight can be harnessed in the generation of solar power, solar water heating, drying foodstuffs and so forth.

Opening Areas to Incoming Energy

An area of a site can be opened up or cleared to allow a natural energy to come into the system more easily.

Sunlight is one of the elements of nature that we might want to increase in our design. If we have structures or trees blocking the light reaching our Zone 1 kitchen garden for example, rather than relocate the garden, we can clear the area to allow more light in. Where places are too shaded, we can thin out trees or branches to increase productivity from our available space.

Similarly, we can clear an area to create a view of a pleasant outside area. If we have potentially stunning views of mountains, lakes, forests or simply an inspiring piece of Nature from the home, we will want to clear any objects obstructing the view to take advantage of such a positive feature in our site design.

Mapping Sectors

To map out how these wild energies interact with our site, we can use a sector diagram.

Each sector indicates one of the external energies discussed above, and is usually represented as a wedge shape, like a slice of a pie, radiating our from the centre of activity, Zone 0, the home, but it can be any other structure of central focus if necessary.

The diagram below shows a sector diagram for the southern hemisphere, with the sun to the north. The sun paths are shown, as well as other key energies. Note that the Zone 5 wildlife corridor extends into Zone 1 in this example, to show demonstrate the flexibility of zone design.

So, we can use sector planning in our design to manage the incoming wild energies moving through our site.

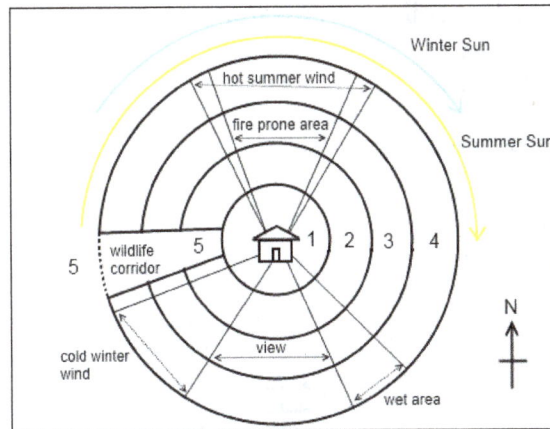

By the strategic placement of elements in our design, we can block, channel, or open up access to these natural energies. to optimise the use of energy in our site.

Together, zone planning and sector planning cover the management of energy inside the site and external energies flowing through the site respectively. Once we have completed our zone and sector analysis, we can then consider one more factor in efficient energy planning, that is the concept of Slope.

Slope

When the site we are designing is on uneven ground, with, slopes, hills and valleys, further design considerations need to be taken into account because the contour of the land has a pronounced effect on the flow of energy in the system.

With any slope, gravity will move things from the highest point to the lowest, and we can take advantage of the work performed by gravitational force to make our system more energy efficient. The main emphasis of designing with slope is efficient energy flow.

Water

Water naturally flows from the highest point to the lower, so slope can be used to move water downhill by gravity.

- By situating water storage such as tanks or dams uphill on the higher points of a site, gravity alone can supply a flow of water without needing additional energy to power a pump.

- Gravel pits with reeds growing in them as a wastewater treatment system located downhill from the house, waste water (greywater) can flow under gravity into the reed beds, where it is cleaned, and then directed into a pond situated further down the slope.

- As an alternative to using a reed bed wastewater treatment system, the wastewater from the kitchen, bathroom and laundry can be directed to an orchard that is located downhill from the house.

- Structures that have a large roof areas for rainwater harvesting, such as sheds, workshops and other such buildings that people don't live in can be located uphill from the house, to capture rainwater, which is collected in water tanks located next to these structures, and fed via gravity to the house.

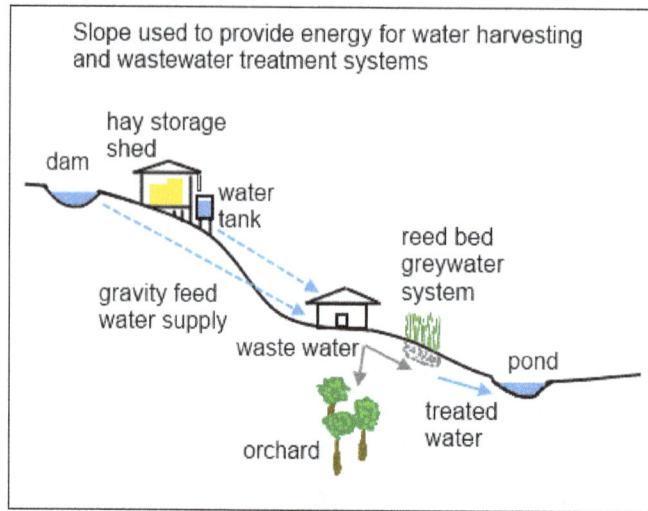

Materials

The movement of resources and materials from the high areas to the lower ones uses less energy than moving them across level ground, and considerable less energy than moving uphill.

By locating access roads uphill of the house, less energy is expended delivering any materials to the site.

Growing timber for firewood or plants for mulch uphill from the house makes it easier to bring the material back to where it will be used because the load is carried downhill.

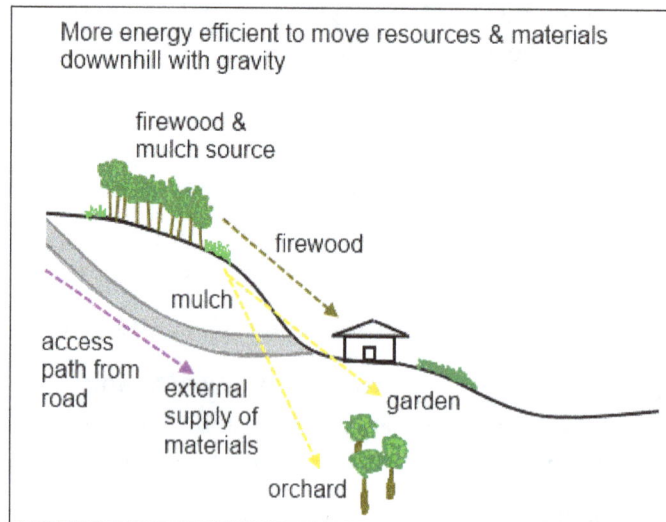

Heat

Heat behaves in an opposite manner to water, as heat rises upwards, as does warm air. Conversely, cold air, being more dense, sinks and flows downward.

Dams and bodies of water situated downslope can reflect heat, as well as act as a thermal mass, heating up during the day, and releasing the heat at night. Since heat rises, the heat emanated will rise upwards and warm the upslope area.

Similarly, we can place plantings of tall trees on a slope to retain heat, to warm the incoming cold night air that flows down the slope. When the warm air moves through the forested area, it will have warmed up as much as it can, and on reaching a plateau at the end of the forest, all the warm air will begin to rise, creating a thermal belt, which will be warmer than the surrounding area. If we place a house in this thermal belt, it will be warmed naturally.

The keypoint referred to in the diagram below is the flat area or plateau that lies between the convex slope above it and the concave slope below it.

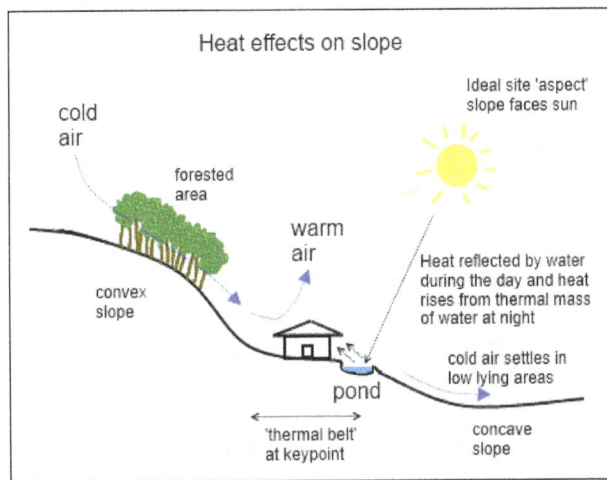

We can also take advantage of the fact that hot air or water rises to set up the collection point downhill and use the energy above the collection point. If we set up a solar water heater downhill, the hot water will naturally rise upwards by convection, and the hot water can be accesses from an elevated tank. This is the principle by which thermosiphons work.

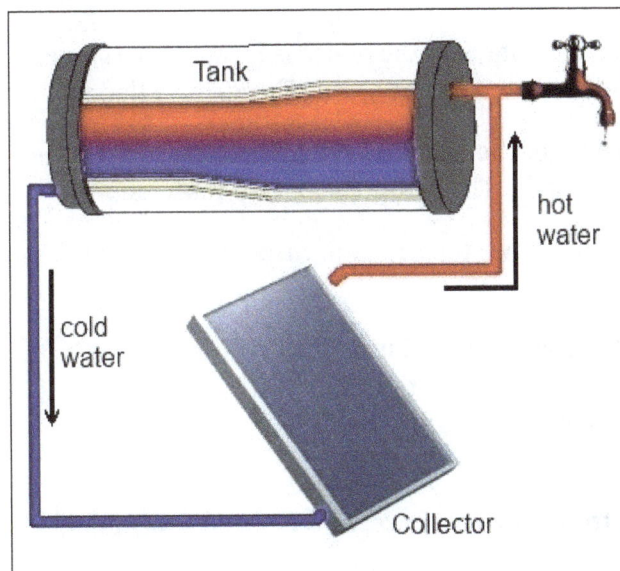

A solar hot water heater is basically a thermosiphon, a passive heat exchanger that works by convection to circulate water without a pump.

The water starts to move when the water inside the collector is heated by the sun. It expands, becomes less dense, and therefore lighter, and rises to float above the cooler, denser, heavier cold water.

As the convection moves the hot water upwards, out of the collector, cold water flows by gravity into the collector where it in turn is warmed up.

Erosion Control

Forested steep slopes not only warm the cool night air to create a thermal belt as described previously, but they also help control soil erosion. When water runs downhill, it will carve its own watercourses and gullies, washing away the soil in the process. Trees, vegetation and ground covers absorb the flow of the water, and by creating a buffer between the flowing water and the soil, they control the problem of soil erosion.

Water flows the fastest straight down a slope, and the effects of erosion will be most pronounced when water has the most direct path down a slope. Additionally, when water flows fast down a slope, very little of it is absorbed into the soil.

By digging trenches on the contours of the slope (contour trenches or swales), the flow of water can be slowed down, and diverted sideways on its downhill journey, to allow it to soak into the soil.

Likewise, when constructing paths, tracks and fences, it is best to have then run along the contours of the site, and not downhill, as downhill running paths will create significant soil erosion, because there are no ground cover plants protecting the soil on a cleared path. Fences holding livestock will become tracks as livestock walk the fence-line day after day, so we also avoid running these straight downhill if possible.

Fire Control

In areas where the ground is not flat, and the house is located on a slope, the biggest danger is from fires running from the downhill area up the slope. These are called upslope fires.

The steeper the slope. the higher the risk. The speed and intensity of the fires doubles for every 10 degree increase in the slope angle. This happens for two reasons:

- The angle of the slope allows the fire to dry the material uphill, making it more flammable when the fire reaches it.

- The updraught effect: When a fire burns, the heat creates an 'updraught', hot air rises fast, and the fire pulls in more oxygen-rich air from lower downhill to feed it. The more air that feeds in, the hotter it gets, and the fiercer it burns.

As a consequence, the worst places to site a house is on sharp ridgetops or hilltops. The house is exposed from all sides to the threat of fire, and fire will race quickly up the slope to reach the house.

Another risky spot is the lee side of a hill, that is, the side of the hill sheltered from the wind. As the wind blows over the crest or top of the hill, it creates a low pressure area on the lee side, which creates a lot of air movement. During a fire, this powerful air movement can drive a fire cyclone, which will be burning directly over the house.

To reduce the risk of fire, houses need to be sited:

- Away from the tops of hills or ridges.

- Preferably on downslope plateaus (level areas).

- If the house is located on a slope of a hillside, excavate a shelf, a flat area, and locate the house on the shelf, well back from the edge to protect it from radiant heat coming from the downhill area.

- If excavating a shelf, build a pond as a firebreak or an earthbank to protect the house from radiant heat.

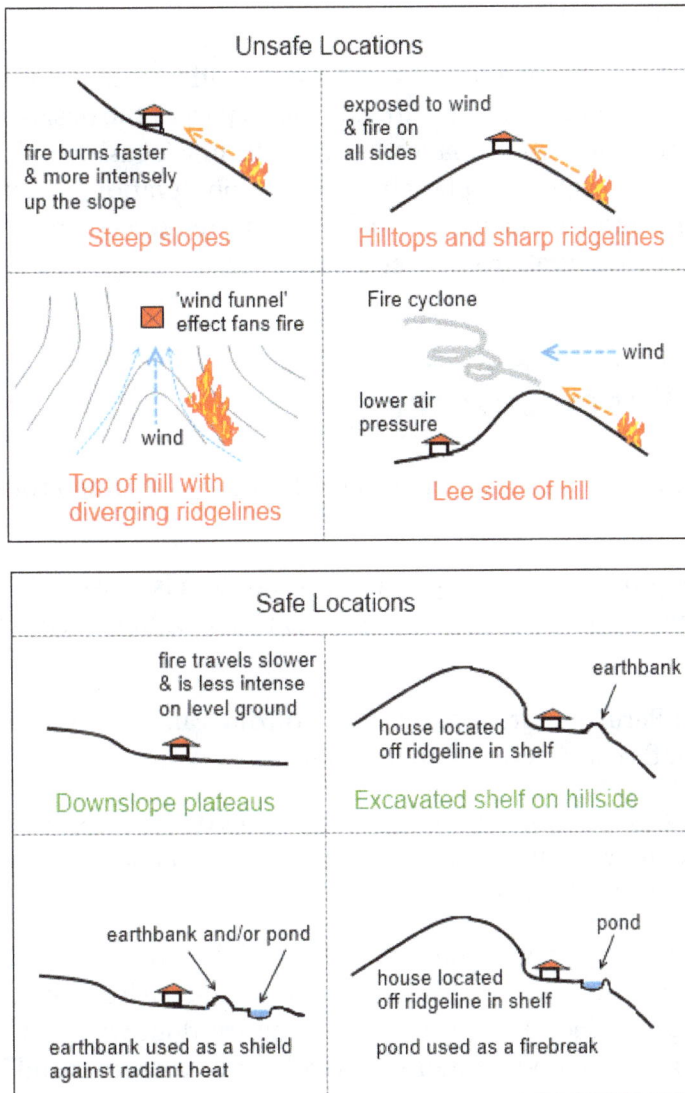

Aspect

The aspect of a slope is simply the orientation of the slope, the direction it faces.

A sun facing slope (facing north in the southern hemisphere, facing south in the northern hemisphere) is the ideal we aim for.

When the slope faces the sun. the site is able to receive the maximum amount of sunlight possible, which means the house, trees and gardens can make the best use of available light, and there will be less issues with the damaging effects of frost. When creating a site design, if we:

- Determine the zones within the site to optimise the distances where elements are located,
- Analyse our sectors carefully to account for all the 'wild energies' moving through our site, and locate design elements to harness or reduce them as necessary,
- Assess the sun angle and slope to gain maximum benefit from them.

Then we will have a fairly sound and potentially successful design in terms of making the most efficient use of energy for our site.

It is a fairly simple and straightforward exercise to systematically step through each of the areas covered under Zone, Sector and Slope, and attend to each part of the design as a separate task. By breaking up even the largest site into smaller sections, it's much easier to design. Dividing up the site into zones does this for us, sector planning involves observation of Nature to see where the elements of nature come into our site design, and slope is a really a creative exercise where we see how much free energy you can grab from what nature offers.

Benefits of Permaculture

Permaculture is an attractive option for landowners because of its numerous benefits. Just some of the benefits include:

- Less waste: Everything within the permaculture system is utilized; like using garden waste as fertilizer. The use of waste and by-products is one of the reasons why permaculture is sustainable.

- Saves on water: Permaculture also involves utilizing rainwater and wastewater, making it more efficient and cost effective than most farms.

- Economically feasible: It is cost efficient since you don't need to use pesticides, and most permaculture systems require less maintenance. All you need to do is water the crops and do mulching from time to time.

Permaculture helps your property withstand the effects of climate change and continue to be productive. It's also an agricultural practice that's healthier to the environment, allowing for sustainable production for longer periods of time. And permaculture, done well, can be more efficient than a regular farm, allowing you to produce more crop with fewer resources and less maintenance.

Applications of Permaculture

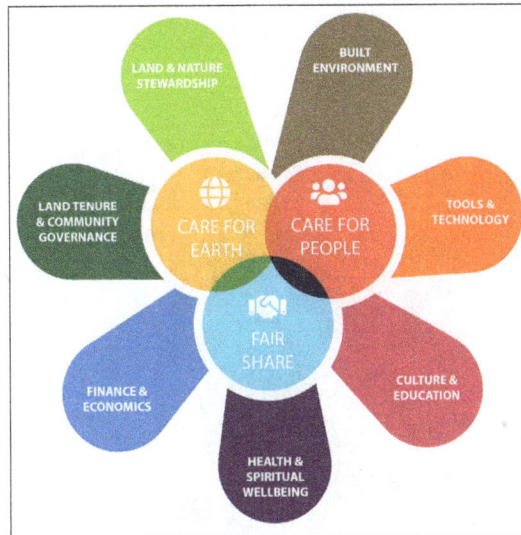

The permaculture flower.

Permaculture incorporates knowledge from many fields of study in its designs and practices, and it can be utilized by and positively impact these fields. These areas of impact can be seen in Holmgren's Permaculture Flower depicted in figure. In this model, the permaculture ethics and principles are centrally located and are depicted as core to the practice, while petals depict areas in which permaculture designs and techniques may be implemented. This particular diagram helps depict permaculture as a natural and social science that is about more than agricultural practices. Fields in which permaculture design can be utilized and incorporated include the building, technology, education, health and spiritual wellbeing, finance and economics, land tenure and community governance, and land and nature stewardship sectors. In many instances, permaculture design may be implemented in these sectors through the use of cooperative models of teaching, ownership and governance. Existing models include credit unions, grocery cooperatives and Waldorf school education (a non-sectarian and non-denominational pedagogical method based on self-governance that recognizes all world cultures and religions as equal and important).

References

- What-is-permaculture: permaculturenews.org, Retrieved 13 June, 2019

- Principles-of-vegan-permaculture: veganaustralia.org.au, Retrieved 28 August, 2019

- Permaculture-Scientific-Evidence-of-Principles-for-the-Agroecological-Design-of-Farming-Systems-327568666: researchgate.net, Retrieved 27 April, 2019

- 4-zones-and-sectors-efficient-energy-planning, permaculture-design-principles: deepgreenpermaculture.com, Retrieved 09 July, 2019

- Different-types-and-benefits-of-permaculture, new-and-views: ruraldesign.co.nz, Retrieved 17 January, 2019

Concepts of Permaculture

2

- **Assisted Natural Regeneration**

- **Polyculture**

- **Conservation Agriculture**

- **Holistic Management**

- **Conservation Grazing**

- **Regenerative Agriculture**

- **Ecosynthesis**

- **Compost Heater**

- **Rotational Grazing**

- **Natural Farming**

- **Synergistic Gardening**

Permaculture comprises of a wide range of concepts which are important in various domains. Assisted natural regeneration, polyculture, conservation agriculture, holistic management, conservation grazing, regenerative agriculture, rotational grazing, etc. are some of the concepts. This chapter discusses in detail these concepts related to permaculture.

Assisted Natural Regeneration

ANR is a method for enhancing the establishment of secondary forest from degraded grassland and shrub vegetation by protecting and nurturing the mother trees and their wildlings inherently present in the area. ANR aims to accelerate, rather than replace, natural successional processes by removing or reducing barriers to natural forest regeneration such as soil degradation, competition with weedy species, and recurring disturbances (e.g., fire, grazing, and wood harvesting). Seedlings are, in particular, protected from undergrowth and extremely flammable plants such as Imperata grass. In addition to protection efforts, new trees are planted when needed or wanted (enrichment planting). With ANR, forests grow faster than they would naturally.

The Benefits of using ANR

ANR provides a range of benefits. ANR is considered to:

- Be a cost efficient way of regenerating forest,

- Provide job opportunities for communities,

- Contribute to strengthening biodiversity,

- Provide hunting areas,

- Increase carbon sequestration and carbon sinks which contribute to climate change mitigation.

Polyculture

Polyculture is a form of agriculture in which more than one species is grown at the same time and place in imitation of the diversity of natural ecosystems. Polyculture is the opposite of monoculture, in which only members of one plant or animal species are cultivated together. Polyculture has traditionally been the most prevalent form of agriculture in most parts of the world and is growing in popularity today due to its environmental and health benefits. There are many types of polyculture including annual polycultures such as intercropping and cover cropping, permaculture, and integrated aquaculture. Polyculture is advantageous because of its ability to control pests, weeds, and disease without major chemical inputs. As such, polyculture is considered a sustainable form of agriculture. However, issues with crop yield and biological competition have caused many modern major industrial food producers to continue to rely on monoculture instead.

Polyculture has traditionally been the most prevalent form of agriculture. A well-known example of historic polyculture is the intercropping of maize, beans, and squash plants in a group often referred to as "the three sisters". In this combination, the maize provides a structure for the bean to grow on, the bean provides nitrogen for all of the plants, while the squash suppresses weeds on the ground. This crop mixture can be traced back several thousand years ago to civilizations in Latin America and Africa and is representative of how species in polycultures sustain each other and minimize the need for human intervention. Integrated aquaculture, or the growing of seafood and plants together, has been common in parts of Eastern Asia for several thousand years as well. In China and Japan, for example, fish and shrimp have historically been grown in ponds with rice and seaweed. Other countries where polyculture has traditionally been a substantial part of agricultural and continues to be so today include those in the Himalayan region, Eastern Asia, South America, and Africa.

Because of the development of pesticides, herbicides, and fertilizers, monoculture became the predominant form of agriculture in the 1950's. The prevalence of polyculture declined greatly in popularity at that time in more economically developed countries where it was deemed to produce less yield while requiring more labor. Polyculture farming has not disappeared entirely though as traditional polyculture systems continue to be an essential part of the food production system today. Around 15% to 20% of the world's agriculture is estimated as relying on traditional polyculture systems. The majority of Latin American farmers continue to intercrop their maize, beans, and squash. Due to climate change, polyculture is returning in popularity in more developed countries as well as food producers seek to reduce their environmental and health impacts.

Common Practice

The kinds of plants that are grown, their spatial distribution, and the time that they spend growing together determines the specific type of polyculture that is implemented. There is no limit for the types of plants or animals that can be grown together to form a polyculture. The time overlaps between plants can be asymmetrical as well, with one plant depending on the other for longer than is reciprocated, often due to differences in life spans.

Annual Polycultures

Intercropping

When more than two crops are grown in complete spatial and temporal overlap with each other, it is referred to as intercropping. Intercropping is particularly useful in plots with limited land availability. Legumes are one of the most commonly intercropped crops, specifically legume-cereal mixtures. Legumes fix atmospheric nitrogen into the soil so that it is available for consumption by other plants in a process known as nitrogen fixation. The presence of legumes consequently eliminates the need for man-made nitrogen fertilizers in intercrops.

Cover Cropping

When a crop is grown alongside another plant that is not a crop, the combination is referred to as cover cropping. If the non-crop plant is a weed, the combination is called a weedy culture. Grasses and legumes are the most common cover crops. Cover crops are greatly beneficial as they can help prevent soil erosion, physically suppress weeds, improve surface water retention, and, in the case of legumes, provide nitrogen compounds as well.

Strip Cropping

Strip cropping is a form of polyculture that involves growing different plants in alternating rows. While strip cropping does not involve the complete intermixing of plant species, it still provides many of the same benefits such as preventing soil erosion and aiding with nutrient cycling.

Permaculture

A coffee farm in Colombia where coffee plants are grown under several tree species in imitation of natural ecosystems. Trees provide valuable resources for the coffee plants such as shade, nutrients, and soil structure.

Permacultures include polycultures of perennial plants such as legume-grass mixtures and wild-flower mixtures, popular in Europe and more temperate climates. These increase soil fertility through nitrogen fixation, decrease soil erosion, regulate water consumption, and decrease the need for tillage thereby conserving soil nutrients. Permacultures require even less human intervention than other forms of polyculture because of lower harvest and tillage rates.

In many Latin American countries, agroforestry is a popular form of permaculture as well where trees and crops are grown together.Trees provide shade for the crops alongside organic matter and nutrients when they shed their leaves or fruits. The elaborate root systems of trees also help prevent soil erosion and increase the presence of microbes in the soil. In addition to benefiting crops, trees act as commodities themselves for use in paper, medicine, firewood, etc. Growing coffee plants alongside other tree species in Mexico is a common practice of agroforestry.

Coffee is a shade-loving crop, and is traditionally shade-grown. In India, it is often grown under a natural forest canopy, replacing the shrub layer. A different polyculture system is used for coffee in Mexico, where the Coffea bushes are grown under leguminous trees in the genus Inga.

One approach to sustainability is to develop polyculture systems by breeding perennial crop varieties of traditional annual arable crops. Perennial crops require less tillage and often have longer roots than annual varieties, helping to reduce soil erosion and tolerate drought. Such varieties are being developed for rice, wheat, sorghum, pigeon pea, barley, and sunflowers. If these can be combined in polyculture with a leguminous cover crop such as alfalfa, fixation of nitrogen will be added to the system, reducing the need for fertilizer and pesticides.

Ducks with free access to rice paddies provide an additional source of income, eat weeds that would restrict rice growth, and manure the fields, reducing the need for fertilizer.

Some traditional systems have combined polyculture with sustainability. In South-East Asia, rice-fish systems on rice paddies have raised freshwater fish as well as rice, producing a valuable additional crop and reducing eutrophication of neighbouring rivers. A variant in Indonesia combines rice, fish, ducks and water fern for a resilient and productive permaculture system; the ducks eat the weeds that would otherwise limit rice growth, saving labour and herbicides, while the duck and fish manure substitute for fertilizer.

Integrated Aquaculture

Integrated aquaculture is a form of aquaculture in which cultures of fish or shrimp are grown together with seaweed, shellfish, or micro-algae. Mono-species aquaculture, a form of aquaculture where only members of the species are grown together, poses several problems for farmers and the environment. The harvesting of seaweed crops in monocultures, for example, releases nitrates into the water and can lead to severe eutrophication as has occurred in the Venice Lagoon. In terms of seafood growth, the greatest problem in monocultures is the high cost of feed, which accounts for about half of production costs. However, more than half of seafood feed is shown to go to waste and can lead to further problems with excess nitrogen release and eutrophication or algal blooms of freshwater. Many technological approaches to lowering these harmful environmental effects such as bacterial bio-filters have proved to consume high levels of energy and be economically costly.

As such, many farmers have transitioned towards integrated aquaculture. In integrated aquaculture farms, plants serve a dual purpose, acting as food for the sea animals and as a water filtration device for the surrounding environment by absorbing nitrates and excess oxygen. Nutrients can be recycled between plants and animals, reducing the need for chemical nutrient supplements. Plants that are grown alongside seafood such as seaweed often hold significant commercial value by themselves, so incorporating them into already existing seafood monocultures increases economic value.

Functions

Pest Management

Pests are less predominant in polycultures than monocultures due to crop diversity. The lack of concentration of a single crop makes polycultures less appealing to pests who have a strong preference towards a specific crop. These specialized pests will often have more difficulty locating a favorable host plant inside of a polyculture than in a monoculture. If a pest has more generalized preferences, it will leave more quickly to other plants in the polyculture and as such have a lesser effect on any one plant. When pests are present in the nearby area, polycultures consequently experience lower yield loss than monocultures do in a theory known as the associational resistance hypothesis. Because polycultures mimic naturally diverse ecosystems, general pests are less likely to distinguish between polycultures and the surrounding environment as well. As such, pests travel more freely between the two environments, and have a relative smaller in presence in polycultures to begin with.

Because of the diversity of plants in a polyculture, natural enemies, or predators, of pests are often attracted to the polyculture alongside pests themselves as well. These natural enemies help suppress pest populations while doing no harm to the plants themselves.

Disease Control

Plant diseases are less predominant in polycultures than monocultures. The disease-diversity hypothesis states that a greater diversity of plants leads to a decreased severity of disease. Because different plants are susceptible to different diseases, if a disease negatively impacts one crop, it will not necessarily spread to another and so the overall impact is contained. However, the type of disease and the susceptibility of the specific plants inside the polyculture to a particular disease can vary greatly.

Weed Management

Both the density of crops and the diversity affect weed growth in polycultures. Having a greater density of plants reduces the available water, sunlight, and nutrient concentrations in the environment. Such a reduction is heightened with greater crop diversity as more potential resources are fully utilized. This level of competition makes polycultures particularly inhospitable to weeds.

When they do grow, weeds can help polycultures, assisting in pest management by attracting natural enemies of pests. They can also act as hosts to arthropods that are beneficial to other plants in the polyculture.

Advantages

Sustainability

Applying pesticides to crops in a monoculture: some of these pesticides will likely end up in water sources and the atmosphere where they can pose serious health and environmental harm. Polycultures attract fewer pests than monocultures and experience less yield loss even when pests are present, often eliminating the need for pesticides.

Because polycultures use methods of pest, disease, and weed control that do not rely on human intervention, they do not release pesticides into the environment. Fertilizer use is reduced as well, as diverse plants more fully share and use all available soil and atmospheric nutrients. As such, environmental impacts such as eutrophication of fresh water or the presence of excess atmospheric nitrogen are greatly reduced.

Other negative impacts of modern agriculture are similarly reduced. Excessive tillage occurs in most modern agricultural practices, but removes essential microbes and nutrients from the soil that are conserved in polycultures, especially permacultures. Because polyculture relies on natural systems of crop maintenance, farmers save money on machinery. By growing multiple plants or animals together in the same space, agricultural land, a critical resource worldwide, taking up 40% of the world's land area, is used more productively.

Polyculture increases local biodiversity. Increasing crop diversity can increase pollination in near-by environments, as diverse plants attract a broader array of pollinators. This is one example of reconciliation ecology, or accommodating biodiversity within human landscapes. This may also form part of a biological pest control program.

Human Health

The chemicals used in monoculture food production can be directly harmful to human health when released into the environment. Nitrogen is a chemical found in especially high concentrations in fertilizers. Nitrates from these fertilizers often become integrated in water sources due to agricultural runoff. The consumption of nitrates at high doses has been shown to lead to methemoglobinemia in infants.

Many of the crops consumed today are calorie-rich crops as well which can lead to illnesses such as obesity, hypertension, and type II diabetes. Because it encourages plant diversity, polyculture can help increase diet diversity by incorporating non-traditional foods into agriculture and people's diets.

Effectiveness

The effects of interspecific competition and intraspecific competition can cause great damage to plants in certain polycultures. In order for a polyculture to be effective, the diverse species that are a part of it must have distinct biological needs such as absorbing different nutrients or requiring different amounts of sunlight as stated by the competitive exclusion principle. Due to the large number of plant species that are cultivated by humans, finding and testing combinations of plants where interspecific and intraspecific competition do not significantly negatively affect the individual plants is extremely difficult. As such, for crops where historic polycultures do not exist, such a multiplicity makes the creation of new polycultures a significant issue.

Crop yield is also an issue in polycultures. While a polyculture produces more biomass overall than a monoculture, individual crops inside of the polyculture are not as prevalent. When there is a focal crop whose cultivation is especially important for a society a lower yield for a certain crop may pose food availability issues.

Similarly, while diseases and pests affect a polyculture less as a group, they do not necessarily have a decreased effect on a focal crop. If targeted by a specialized pest or disease, a focal crop in a polyculture will likely experience the same yield loss as its monoculture counterpart.

Polyculture also often requires more labor.

Conservation Agriculture

Conservation agriculture (CA) can be defined by a statement given by the Food and Agriculture Organization of the United Nations as "a concept for resource-saving agricultural crop production that strives to achieve acceptable profits together with high and sustained production levels while concurrently conserving the environment".

Agriculture according to the New Standard Encyclopedia is "one of the most important sectors in the economies of most nations". At the same time conservation is the use of resources in a manner that safely maintains a resource that can be used by humans. Conservation has become critical because the global population has increased over the years and more food needs to be produced every year. Sometimes referred to as "agricultural environmental management", conservation agriculture may be sanctioned and funded through conservation programs promulgated through agricultural legislation, such as the U.S. Farm Bill.

Key Principles

The Food and Agriculture Organization of the United Nations (FAO) has determined that CA has three key principles that producers (farmers) can proceed through in the process of CA. These three principles outline what conservationists and producers believe can be done to conserve what we use for a longer period of time.

The first key principle in CA (Conservation Agriculture) is practicing minimum soil disturbance which is essential to maintaining minerals within the soil, stopping erosion, and preventing water loss from occurring within the soil. In the past agriculture has looked at soil tillage as a main process in the introduction of new crops to an area. It was believed that tilling the soil would increase fertility within the soil through mineralization that takes place in the soil. Also tilling of soil can cause severe erosion and crusting which leads to a decrease in soil fertility. Today tillage is seen as destroying organic matter that can be found within the soil cover. No-till farming has caught on as a process that can save soil organic levels for a longer period and still allow the soil to be productive for longer periods. Additionally, the process of tilling can increase time and labor for producing that crop. Minimum soil disturbance also reduce destruction of soil micro and macro-organism habitats that is common in conventional ploughing practices.

When no-till practices are followed, the producer sees a reduction in production cost for a certain crop. Tillage of the ground requires more money in order to fuel tractors or to provide feed for the animals pulling the plough. The producer sees a reduction in labor because he or she does not have to be in the fields as long as a conventional farmer.

The second key principle in CA is much like the first in dealing with protecting the soil. The principle of managing the top soil to create a permanent organic soil cover can allow for growth of organisms within the soil structure. This growth will break down the mulch that is left on the soil surface. The breaking down of this mulch will produce a high organic matter level which will act as a fertilizer for the soil surface. If CA practices were used done for many years and enough organic matter was being built up at the surface, then a layer of mulch would start to form. This layer helps prevent soil erosion from taking place and ruining the soil's profile or layout. The presence of mulching also reduce the velocity of runoff and the impact of rain drops thus reducing soil erosion and runoff.

According to the article "The role of conservation agriculture and sustainable agriculture", the layer of mulch that is built up over time will become like a buffer zone between soil and mulch and this will help reduce wind and water erosion. With this comes the protection of the soil's surface when rain falls on the ground. Land that is not protected by a layer of mulch is left open to the elements. This type of ground cover also helps keep the temperature and moisture levels of the soil at a higher level rather than if it was tilled every year.

The third principle is the practicing diverse crop rotations or crop interactions. According to an article published in the Physiological Transactions of the Royal Society called "The role of conservation agriculture and sustainable agriculture," crop rotation can be used best as a disease control against other preferred crops. This process will not allow pests such as insects and weeds to be set into a rotation with specific crops. Rotational crops will act as a natural insecticide and herbicide against specific crops. Not allowing insects or weeds to establish a pattern will help to eliminate problems with yield reduction and infestations within fields. Crop rotation can also help build up soil infrastructure. Establishing crops in a rotation allows for an extensive buildup of rooting zones which will allow for better water infiltration.

Organic molecules in the soil break down into phosphates, nitrates and other beneficial elements which are thus better absorbed by plants. Plowing increases the amount of oxygen in the soil and increases the aerobic processes, hastening the breakdown of organic material. Thus more nutrients are available for the next crop but, at the same time, the soil is depleted more quickly of its nutrient reserves.

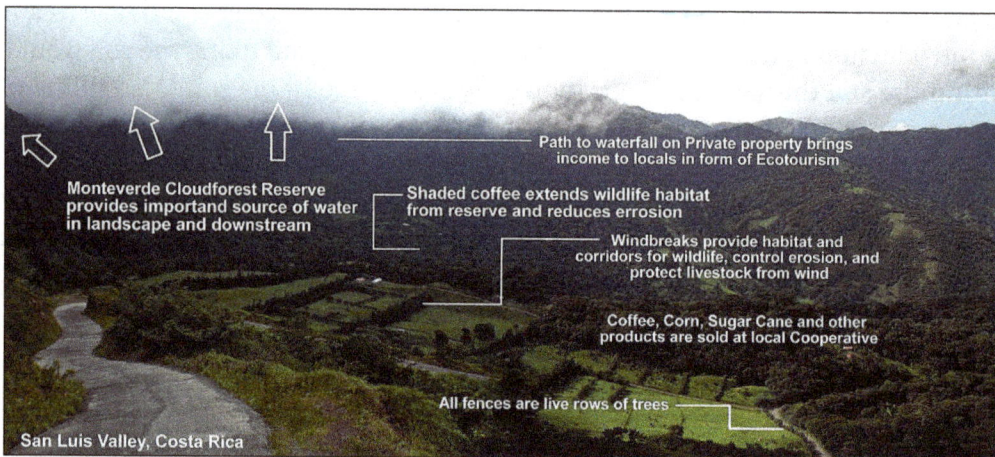

Conservation- or eco-agriculture involves multiple elements to protect wildlife.

In conservation agriculture there are many examples that can be looked towards as a way of farming and at the same time conserving. These practices are well known by most producers. The process of no-till is one that follows the first principle of CA, causing minimal mechanical soil disturbance. No-till also brings other benefits to the producer. According to the FAO, tillage is one of the most "energy consuming" processes that can be used: It requires a lot of labor, time, and fuel to till. Producers can save 30% to 40% of time and labor by practicing the no-till process.

Besides conserving the soil, there are other examples of how CA is used. According to an article in Science called "Farming and the Fate of Wild Nature" there are two more kinds of CA. The practice of wildlife-friendly farming and land sparing are ideas for producers who are looking to practice better conservation towards biodiversity.

Wildlife-friendly Farming

Wildlife-friendly farming is a practice of setting aside land that will not be developed by the producer (farmer). This land will be set aside so that biodiversity has a chance to establish itself in areas with agricultural fields. At the same time, the producer is attempting to lower the amount of fertilizer and pesticides used on the fields so that organisms and microbial activity have a chance to establish themselves in the soil and habitat. But as in all systems, not all can be perfect. To create a habitat suitable for biodiversity something has to be reduced, and as in this case for agriculture farmers, yields can be reduced. This is where the second idea of land sparing can be looked on as an alternative manner.

Land Sparing

Land sparing is another way that producer and conservationist can be on the same page. Land sparing advocates for the land that is being used for agricultural purposes to continue to produce crops at increased yield. With an increase in yield on all land that is in use, other land can be set aside for conservation and production for biodiversity. Agricultural land stays in production but would have to increase its yield potential to keep up with demand. Land that is not being put into agriculture would be used for conserving biodiversity. In fact, data from the Food and Agriculture Organization shows that between 1961 and 2012, the amount of arable land needed to produce the same amount of food declined by 68 percent worldwide.

Benefits

In the field of CA there are many benefits that both the producer and conservationist can obtain.

On the side of the conservationist, CA can be seen as beneficial because there is an effort to conserve what people use every day. Since agriculture is one of the most destructive forces against biodiversity, CA can change the way humans produce food and energy. With conservation come environmental benefits of CA. These benefits include less erosion possibilities, better water conservation, improvement in air quality due to lower emissions being produced, and a chance for larger biodiversity in a given area.

On the side of the producer and farmer, CA can eventually do all that is done in conventional agriculture, and it can conserve better than conventional agriculture. CA according to Theodor Friedrich, who is a specialist in CA, believes "Farmers like it because it gives them a means of conserving, improving, and making more efficient use of their natural resources". Producers will find that the benefits of CA will come later rather than sooner. Since CA takes time to build up enough organic matter and have soils become their own fertilizer, the process does not start to work overnight. But if producers make it through the first few years of production, results will start to become more satisfactory.

CA is shown to have even higher yields and higher outputs than conventional agriculture once it has been established over long periods. Also, a producer has the benefit of knowing that the soil in which his crops are grown is a renewable resource. According to New Standard Encyclopedia, soils are a renewable resource, which means that whatever is taken out of the soil can be put back over time. As long as good soil upkeep is maintained, the soil will continue to renew itself. This could be very beneficial to a producer who is practicing CA and is looking to keep soils at a productive level for an extended time.

The farmer and producer can use this same land in another way when crops have been harvested. The introduction of grazing livestock to a field that once held crops can be beneficial for the producer and also the field itself. Livestock manure can be used as a natural fertilizer for a producer's field which will then be beneficial for the producer the next year when crops are planted once again. The practice of grazing livestock using CA helps the farmer who raises crops on that field and the farmer who raises the livestock that graze off that field. Livestock produce compost or manure which are a great help in generating soil fertility. The practices of CA and grazing livestock on a field for many years can allow for better yields in the following years as long as these practices continue to be followed.

The FAO believes that there are three major benefits from CA:

- Within fields that are controlled by CA the producer will see an increase in organic matter.

- Increase in water conservation due to the layer of organic matter and ground cover to help eliminate transportation and access runoff.

- Improvement of soil structure and rooting zone.

Future Development

As in any other business, producers and conservationists are always looking towards the future. In this case CA is a very important process to be looked at for future generation. There are many organizations that have been created to help educate and inform producers and conservationists in the world of CA. These organizations can help to inform, conduct research, and buy land in order to preserve animals and plants.

Another way in which CA is looking to the future is through prevention. According to the European Journal of Agronomy producers are looking for ways to reduce leaching problems within their fields. These producers are using the same principles within CA, in that they are leaving cover over their fields in order to save fields from erosion and leaching of chemicals. Processes and studies like this are allowing for a better understanding of how to conserve what we are using and finding ways to put back something that may have been lost before.

In the same journal article is presented another way in which producers and conservationists are looking towards the future. Circulation of plant nutrients can be a vital part for conserving the future. An example of this would be the use of animal manure. This process has been used for quite some time now, but the future is looking towards ways to handle and conserve nutrients within manure for a longer time. But besides animal waste, food and urban waste are also being looked towards as a way to use growth within CA. Turning these products from waste to being used to grow crops and improve yields is something that would be beneficial for conservationists and producers.

Agri-environment Schemes

In 1992, 'agri-environment schemes' became compulsory for all European Union Member States. In the following years the main purpose of these schemes changed slightly. Initially, they sought to protect threatened habitats, but gradually shifted their focus to the prevention of the loss of wildlife from agricultural landscapes. Most recently, the schemes are placing more emphasis on improving the services that the land can provide to humans (e.g. pollination). Overall, farmers involved in the scheme aim to practice environmentally friendlier farming techniques such as: reducing the use of pesticides, managing or altering their land to increase more wildlife friendly habitats (e.g. increasing areas of trees and bushes), reducing irrigation, conserving soil, and organic farming.

As the changes in practices that ensure the protection of the environment are costly to farmers, the EU developed agri-environment schemes to financially compensate individual farmers for applying these changes and therefore increased the implementation of conservation agriculture. The schemes are voluntary for farmers. Once joined, they commit to a minimum of five years during which they have to adopt various sustainable farming techniques. According to the Euro-stat website, in 2009 the agricultural area enrolled in agri-environment schemes covered 38.5 million hectares (20.9% of agricultural land in the 27 member states of the EU at the time) (Agri-environmental indicator 2015). The European Commission spent a total of €3.23 billion on agri-environment schemes in 2012, significantly exceeding the cost of managing special sites of conservation that year, which came to a total of €39.6 million. There are two main types of agri-environment schemes which have shown different outcomes. Out-of-production schemes tend to be used in extensive farming practices (where the farming land is widespread and less intensive farming is practiced), and focus on improving or setting land aside that will not be used for the production of food, for example, the addition of wildflower strips. In-production schemes (used for a smaller scale, but more intensively farmed land) focus on the sustainable management of arable crops or grassland, for example reduction of pesticides, reduction of grassland mowing, and most commonly, organic farming. In a 2015 review of studies examining the effects of the two schemes, it was found that out-of-production schemes had a higher success rate at enhancing the number of thriving species around the land. The reason behind this is thought to be the scheme's focus on enhancing specific species by providing them with more unaltered habitats, which results in more food resources for the specific species. On the other hand, in-production schemes attempt to enhance the quality of the land in general, and are thus less species specific. Based on the findings, the reviewers suggest that schemes which more specifically target the declining groups of species, may be more effective. The findings and the targets will be implemented between 2015 and 2020, so that by 2025, the effectiveness of these schemes can be re-assessed and will have increased significantly.

Problems

As much as conservation agriculture can benefit the world, there are some problems that come with it. There are many reasons why conservation agriculture cannot always be a win-win situation.

There are not enough people who can financially turn from conventional farming to conservation. The process of CA takes time; when a producer first becomes a conservationist, the results can be a financial loss to them(in most cases, the investment and policy generally exist). CA is based upon establishing an organic layer and producing its own fertilizer and this may take time. It can be many years before a producer will start to see better yields than he/she has had previously. Another financial undertaking is purchasing of new equipment. When starting to use CA, a producer may have to buy new planters or drills in order to produce effectively. These financial tasks are ones that may impact whether or not a producer decides to switch to CA or not.

With the struggle to adapt comes the struggle to make CA grow across the globe. CA has not spread as quickly as most conservationists would like. The reason for this is because there is not enough pressure for producers in places such as North America to change their way of living to a more conservationist outlook. But in the tropics there is more pressure to change to conservation areas because of the limited resources that are available. Places like Europe have also started to catch onto the ideas and principles of CA, but still nothing much is being done to change due to there being a minimal amount of pressure for people to change their ways of living.

With CA comes the idea of producing enough food. With cutting back in fertilizer, not tilling the ground, and other processes comes the responsibility to feed the world. According to the Population Reference Bureau, there were around 6.08 billion people on Earth in the year 2000. By 2050 there will be an estimated 9.1 billion people. With this increase comes the responsibility for producers to increase food supply using the same or less land than we use today. Problems arise in the fact that if CA farms do not produce as much as conventional farms, this leaves the world with less food for more people.

Holistic Management

Holistic management in agriculture is an approach to managing resources that was originally developed by Allan Savory.

Holistic planned grazing is similar to rotational grazing but differs in that it more explicitly recognizes and provides a framework for adapting to the four basic ecosystem processes: the water cycle, the mineral cycle including the carbon cycle, energy flow, and community dynamics (the relationship between organisms in an ecosystem), giving equal importance to livestock production and social welfare. Holistic management has been likened to "a permaculture approach to rangeland management".

Framework

The holistic management decision-making framework uses six key steps to guide the management of resources:

- Define in its entirety what you are managing. No area should be treated as a single-product system. By defining the whole, people are better able to manage. This includes identifying the available resources, including money, that the manager has at his disposal.

- Define what you want now and for the future. Set the objectives, goals and actions needed to produce the quality of life sought, and what the life-nurturing environment must be like to sustain that quality of life far into the future.

- Watch for the earliest indicators of ecosystem health. Identify the ecosystem services that have deep impacts for people in both urban and rural environments, and find a way to easily monitor them. One of the best examples of an early indicator of a poorly functioning environment is patches of bare ground. An indicator of a better functioning environment is newly sprouting diversity of plants and a return or increase of wildlife.

- Don't limit the management tools you use. The eight tools for managing natural resources are money/labor, human creativity, grazing, animal impact, fire, rest, living organisms and science/technology. To be successful you need to use all these tools to the best of your ability.

- Test your decisions with questions that are designed to help ensure all your decisions are socially, environmentally and financially sound for both the short and long term.

- Monitor proactively, before your managed system becomes more imbalanced. This way the manager can take adaptive corrective action quickly, before the ecosystem services are lost. Always assume your plan is less than perfect and use a feedback loop that includes monitoring for the earliest signs of failure, adjusting and re-planning as needed. In other words use a "canary in a coal mine" approach.

Four Principles

Savory stated four key principles of holistic management planned grazing, which he intended to take advantage of the symbiotic relationship between large herds of grazing animals and the grasslands that support them:

- Nature functions as a holistic community with a mutualistic relationship between people, animals and the land. If you remove or change the behavior of any keystone species like the large grazing herds, you have an unexpected and wide-ranging negative impact on other areas of the environment.

- It is absolutely crucial that any agricultural planning system must be flexible enough to adapt to nature's complexity, since all environments are different and have constantly changing local conditions.

- Animal husbandry using domestic species can be used as a substitute for lost keystone species. Thus when managed properly in a way that mimics nature, agriculture can heal the land and even benefit wildlife, while at the same time benefiting people.

- Time and timing is the most important factor when planning land use. Not only is it crucial to understand how long to use the land for agriculture and how long to rest, it is equally important to understand exactly when and where the land is ready for that use and rest.

Beginnings

The idea of holistic planned grazing was developed in the 1960s by Allan Savory, a wildlife biologist in his native Southern Rhodesia. Setting out to understand desertification in the context of the larger environmental movement, and influenced by the work of André Voisin, he hypothesized that the spread of deserts, the loss of wildlife, and the resulting human impoverishment were related to the reduction of the natural herds of large grazing animals and, even more, the changed behavior of the few remaining herds. Savory hypothesized further that livestock could be substituted for natural herds to provide important ecosystem services like nutrient cycling. However, while livestock managers had found that rotational grazing systems can work for livestock management purposes, scientific experiments demonstrated it does not necessarily improve ecological issues such as desertification. As Savory saw it, a more comprehensive framework for the management of grassland systems — an adaptive, holistic management plan — was needed.

Development

In many regions, pastoralism and communal land use are blamed for environmental degradation caused by overgrazing. After years of research and experience, Savory came to understand this assertion was often wrong, and that sometimes removing animals actually made matters worse. This concept is a variation of the trophic cascade, where humans are seen as the top level predator and the cascade follows from there.

Savory developed a management system that he claimed would improve grazing systems. Holistic planned grazing is one of a number of newer grazing management systems that aim to more closely simulate the behavior of natural herds of wildlife and have been shown to improve riparian habitats and water quality over systems that often led to land degradation, and be an effective tool to improve range condition for both livestock and wildlife.

Uses

While originally developed as a tool for range land use and restoring desertified land, the holistic management system can be applied to other areas with multiple complex socioeconomic and environmental factors. One such example is integrated water resources management, which promotes sector integration in development and management of water resources to ensure that water is allocated fairly between different users, maximizing economic and social welfare without compromising the sustainability of vital ecosystems. Another example is mine reclamation. A fourth use of holistic management is in certain forms of no till crop production, intercropping, and permaculture. Holistic management has been acknowledged by the United States Department of Agriculture.

Conservation Grazing

Conservation grazing or targeted grazing is the use of semi-feral or domesticated grazing livestock to maintain and increase the biodiversity of natural or semi-natural grasslands, heathlands, wood pasture, wetlands and many other habitats. Conservation grazing is generally less intensive than practices such as prescribed burning, but still needs to be managed to ensure that overgrazing does not occur. The practice has proven to be beneficial in moderation in restoring and maintaining grassland and heathland ecosystems. The optimal level of grazing will depend on the goal of conservation, and different levels of grazing, alongside other conservation practices, can be used to induce the desired results.

For historic grasslands, grazing animals, herbivores, were a crucial part of the ecosystem. When grazers are removed, historically grazed lands may show a decline in both the density and the diversity of the vegetation. The history of the land may help ecologists and conservationists determine the best approach to a conservation project.

Historic threats to grasslands began with land conversion to crop fields. This shifted to improper land management techniques and more recently to the spread of woody plants due to a lack of management and to climate change.

Conservation Grazing in Practice

Urban ecopastoralism on a Historical Monument, the Citadel of Lille, by Soay sheep.
Their long coats help to disperse plant seeds.

Urban ecopastoralism with sheep and goats in an urban meadow of the
"Bois de la Citadelle" in Lille.

Conservation grazing Longhorn Cattle to manage national
nature reserve at Ruislip Lido.

Grazing maintains an area as a habitat dominated by grasses and small shrubs, largely preventing ecological succession to forest.

Conservation grazing needs to be monitored closely. Overgrazing may cause erosion, habitat destruction, soil compaction, or reduced biodiversity (species richness). Rambo and Faeth found that the use of vertebrates for grazing of an area increased the species richness of plants by decreasing the abundance of dominant species and increasing the richness of rarer species. This may lead to a more open forest canopy and more room for other plant species to emerge.

Grazing Restoration Effect Dependent on Grazer Species

Different grazing species have different effects. For example, elk and horses have a similar grazing frequency to cattle but tend to spread their zone of grazing to cover a greater area, producing a smaller effect on a given area than cattle would. Similarly, cattle have been found to be more useful in the restoration of pastures with low species richness, and sheep were found useful for the re-establishment of neglected fields. The type of area that needs to be restored or maintained will determine the species of grazer ideal for conservation grazing. Dumont et al. found in the use of varied breeds of steers that "traditional breeds appeared slightly less selective than commercial breeds", but did not make a significant difference in biodiversity. In this particular study biodiversity was maintained by the same amount by both breed types.

Effects on Native and Non-native Plant Species

Conservation grazing is a tool used for conserving biodiversity. However, one danger in grazing is the potential for invasive species to be enhanced as well as the native biodiversity. A study by Loeser et al. showed that areas of high intensity grazing and grazer removal increased the biomass of nonnative introduced species. Both showed that an intermediate approach is the best method. The nonnatives did demonstrate that they were not as well adapted to the disturbances, such as drought. This indicated that implementing controlled grazing methods would decrease the abundance of nonnatives in those plots that had not been properly managed.

Effects of grazing can also depend on the individual plant species and its response to grazing. Plants that are adapted to extensive grazing (such as that done by cattle) will respond quicker and more effectively to grazing than native species that have not had to cope with intense grazing pressure in the past. An experiment done by Kimball and Schiffman showed that grazing increased the cover of some native species but did not decrease the cover of nonnative species. The species diversity of the native plants was able to respond to the grazing and increase diversity. The community would become denser than originally with the increased biodiversity. (However, this may have been simply variance in plots due to the fact that the native and nonnative compositions were of different species between the grazed and ungrazed plots).

Effects on Animals

Insects and Butterflies

Degree of grazing has a significant effect on the species richness and abundance of insects in grasslands. Land management in the form of grazing tends to decrease diversity with increased intensity. Kruess and Tscharntke attribute this difference to the increased height of grasses in the ungrazed areas. The study showed that the abundance and diversity of insects (such as butterfly adults, trap-nesting bees and wasps) were increased by increased grass height. However, other insects such as grasshoppers responded better to heterogeneity of the vegetation.

Vertebrates

Grazing can have varied effects on vertebrates. Kuhnert et al. observed that different bird species react in different ways to changes in grazing intensity. Grazing has also been thought to decrease the abundance of vertebrates, such as the prairie dog and the desert tortoise. However, Kazmaier et al. found that moderate grazing by cattle had no effect on the Texas tortoise.

Rabbits have been widely discussed due to their influences on land composition. Bell and Watson found that rabbits show grazing preference for different plant species. This preference can alter the composition of a plant community. In some cases, if the preference is for a non-native, invasive plant, rabbit grazing may benefit the community by reducing non-native abundance and creating room for the native plant species to fill. When rabbits graze in moderation they can create a more complex ecosystem, by creating more variable environments that will allow for more predator-competitor relationships between the various organisms. However, besides the effect on wild vegetation, rabbits destroy crops, compete with other herbivores, and can result in extreme ecological damage. Competition can be direct or indirect. The rabbits may specifically eat the competitions target food or it may inhibit the growth of grasses that other species eat. For example, rabbit grazing in the Netherlands inhibits tall grasses from becoming dominant. This in turn enhances the suitability of the pasture for brent goose. However, they may benefit predators that do better in open areas, because the rabbits reduce the amount of vegetation making it easier for those predators to spot their prey.

Finally, grazing has demonstrated use in clearing dry brush to reduce the fire hazard of drought-stricken areas.

Effect on Ephemeral Wetlands

Ephemeral wetlands degradation and loss of biodiversity had, at one point in time, been blamed on mismanaged grazing of both native and non-native ungulates and other grazers. A study done by Jaymee Marty of The Nature Conservancy examined the effects on the vernal pools formed in California when grazers were removed. The results of the short study showed that areas where grazers were removed had a lower diversity of native grasses, invertebrates and vertebrates in the pools, with an increase in non-native grass abundance and distribution in the area. The study also demonstrated reduced reproduction success of individual species in the area, such as the western spadefoot toad and California tiger salamander. Marty argues that this decrease is due to ecosystems adapting to historical changes in grazers and the effects they have. In other words, the historic ecosystem, theoretically, would have responded positively to the removal of cattle grazing, however, the system has adapted to the European introduced species and now may require them for maintained diversity. In another study performed by Pyke and Marty, measurements showed that on average, vernal ponds on grazed land pooled longer than ungrazed areas and soil was more resistant to water absorption in the grazed areas.

Targeted Grazing

A recent synonym or near-synonym for conservation grazing is "targeted grazing", a term introduced in a 2006 handbook,in distinction to prescribed grazing, which the USDA National Resource Conservation Service was using to describe all managed grazing.Targeted grazing is often used in combination with other techniques such as burning, herbicide applications or land clearing. Targeted grazing can rival traditional herbicide and mechanical control methods for invasive plants from invasive forb to juniper trees, and has been used to reduce fine fuels in fire prone areas.

Principles

The most important skill for developing a targeted grazing program is patience and commitment. However, understanding livestock and plant responses to grazing are critical in developing a targeted grazing program. The program should have a clear statement of the kind of animal, timing and rate of grazing necessary to suppress troublesome plants and maintain a healthy landscape. The grazing application should 1) cause significant damage to the target plants 2) limit damage to desired vegetation and 3) be integrated with other control strategies. First, to cause significant damage to targeted plants requires understanding when the target plant is most susceptible to grazing damage and when they are most palatable to livestock. Target plant palatability depends on the grazing animals inherited and developed plant preferences (i.e. the shape of sheep and goat's mouths make them well suited for eating broad leaf weeds). Goats are also well designed for eating shrubs. Second, target plants often exist in a plant community with many desirable plants. The challenge is to select the correct animal, grazing time and grazing intensity to maximize the impact on the target plant while reducing it on the associated plant community. Finally, management objectives, target plant species, weather, topography, plant physiology, and associated plant communities are among the many variables that can determine treatment type and duration. Well-developed targeted grazing objectives and an adaptive management plan that takes into account other control strategies need to be in place.

Regenerative Agriculture

Regenerative agriculture is a conservation and rehabilitation approach to food and farming systems. It focuses on topsoil regeneration, increasing biodiversity, improving the water cycle, enhancing ecosystem services, supporting biosequestration, increasing resilience to climate change, and strengthening the health and vitality of farm soil. Practices include, recycling as much farm waste as possible, and adding composted material from sources outside the farm.

Biodiversity.

Regenerative agriculture on small farms and gardens is often based on ideologies like permaculture, agroecology, agroforestry, restoration ecology, keyline design and holistic management. Large farms tend to be less ideology driven, and often use "no-till" and "reduced till" practices.

Hoverfly at work.

On a regenerative farm, yield should increase over time. As the topsoil deepens, production may increase and less external compost inputs are required. Actual output is dependent on the nutritional value of the composting materials, and the structure and content of the soil.

Roots

Rodale Institute, Test Garden.

Field Hamois Belgium Luc Viatour.

Regenerative agriculture is based on various agricultural and ecological practices, with a particular emphasis on minimal soil disturbance and the practice of composting. Maynard Murray had similar ideas, using sea minerals. Her work led to innovations in no-till practices, such as slash and mulch in tropical regions. Sheet mulching is a regenerative agriculture practice that smothers weeds and adds nutrients to the soil.

Principles and Practices

Regenerative agriculture is guided by a set of principles and practices.

Principles

Principles include:

- Increase soil fertility.

- Work with whole systems (holistically), not isolated parts, to make changes to specific parts.

- Improve whole agro-ecosystems (soil, water and biodiversity).

- Connect the farm to its larger agro-ecosystem and region.

- Make holistic decisions that express the value of farm contributors.

- Each person and farm is significant.

- Making sure all stakeholders have equitable and reciprocal relationships.

- Being paid or paying others can be with financial, spiritual, social or environmental capital ("multi-capital"). Relationships can be "non-linear" (not reciprocal): if you do not get paid, in the future you can be given other "capital" by unrelated parties.

- Continually grow and evolve individuals, farms and communities.

- Continuously evolve the agro-ecology.

- Agriculture influences the world.

Practices

Practices include but are not limited to:

- Permaculture design.

- Agroforestry.

- Soil food web.

- Properly managed livestock, well-managed grazing, animal integration, and holistically managed grazing.

- STUN (Sheer, Total and Utter Neglect) breeding.

- Keyline Subsoiling.

- Conservation farming, No-Till Farming, minimum tillage, and Pasture Cropping.

- Cover crops & multi-species cover crops.

- Organic Annual Cropping and Crop rotations.

- Compost, Compost Tea, animal manures and Thermal Compost.

- Natural sequence farming.

- Grassfed livestock.

- Polyculture and full-time succession planting of multiple and inter-crop plantings.

- Borders planted for pollinator habitat and other beneficial insects.

- Biochar/Terra Preta.

- Ecological Aquaculture.

- Perennial Crops.

- Silvopasture.

Ecosynthesis

Ecosynthesis is the use of introduced species to fill niches in a disrupted environment, with the aim of increasing the speed of ecological restoration. This decreases the amount of physical damage done in a disrupted landscape. An example is using willow in a stream corridor for sediment and phosphorus capture. It aims to aid ecological restoration which, is the practice of renewing and restoring degraded, damaged, or destroyed ecosystems and habitats in the environment by active human intervention and action. Humans ecosynthesis to make environments more suitable for life, through restoration ecology (introduced species, vegetation mapping, habitat enhancement, remediation and mitigation).

Magnolia Tree, as an introduced Species in a disturbed environment.

Planetary Ecosynthesis

Earth.

A controversial science of applying ecosynthesis to other planets to make them habitable like Earth. This is a futuristic science that has almost no hold on society today due to its large manufacturing costs and level of technology available.

Restoration Ecology

Ecological restoration aims to recreate, initiate, or accelerate the recovery of an ecosystem that has been disturbed.

- Revegetation: The establishment of vegetation on sites where it has been previously lost, often with erosion control as the primary goal.

- Habitat enhancement: The process of increasing the suitability of a site as habitat for some desired species.

- Remediation: Improving an existing ecosystem or creating a new one with the aim of replacing another that has deteriorated or been destroyed.

- Mitigation: Legally mandated remediation for loss of protected species or an ecosystem.

Through restoration ecology humans can help ecosystems that we have either caused harm to or disturbed be brought back to functional state.

Trophic Cascade

A clear example of humans ecosynthesiszing would be through the introduction of a species to cause a trophic cascade, which is the result of indirect effects between nonadjacent trophic levels in a food chain or food web, such as the top predator in a food chain and a plant. The most famous example of a trophic cascade is that of the introduction of wolves to YellowStone National Park, which had extradionary effects to the ecosystem. Yellowstone National Park had a massive population of elk because they had no predators, which caused the local aspen population and other vegetation to significantly decrease in population size. However, the introduction of wolves controlled the elk population and indirectly affected the aspen and other vegetation. This brought the ecosystem back to a sustainable life.

Gray wolf in YellowStone National Park.

Compost Heater

A compost heater (or Biomeiler) is a structure for the energetic use of biomass for the heating of buildings.

The method was developed by Jean Pain in the 1970s. Compost heaters are used primarily for demonstration purposes as small systems for heating a house. Local waste can be converted to energy.

Types

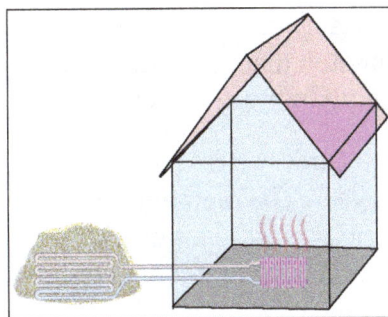

Using the heat of an external compost heap for warming up the inside of a house.

Compost Heap

The traditional compost heater exploits the heat of a large compost heap to warm a house. This type requires a big heap, intertwined with a spiral water hose. The circulating water conducts heat to the building, where it can be fed to a heating circuit.

The heap must contain at least 8,000 liters of biomass to maintain a temperature during the winter.

For this purpose, chipped wood is usually piled up and a water hose is passed through it. A microbiological degradation process generates heat for up to 24 months. The heat produces hot water, which is then fed to a heating circuit. With sufficient oxygen supply, the biomass is degraded by aerobic decomposition. The activity of the microorganisms can be regulated by the moisture content.

Hot Water and Biogas

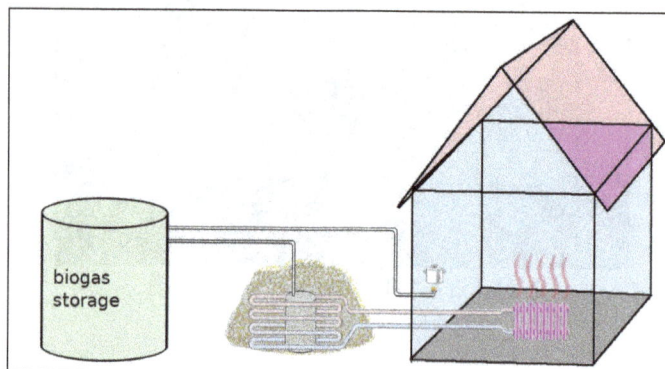

The heat of a compost heap is used, and biogas is produced at the same time.

Pain's 'Biomeiler' combines composting with biogas.

The raw materials of Pain's compost heap were saplings, branches and underbrush. He developed the machines that grind these materials to the proper size. One of his machines, a tractor-driven model, earned fourth prize in the 1978 Grenoble Agricultural Fair. After Pain had ground the raw materials, Pain would construct a heap three metres high and six metres across (10 × 20 feet). The heap weighed approximately 50 tonnes (49 long tons; 55 short tons), and was mounted over a 4 cubic metres (140 cu ft) steel tank. This tank was 3/4 full of compost, which had first been steeped in water for two months. The hermetically sealed tank was connected by tubing to 24 truck tire inner tubes, gathered nearby to collect the methane gas. The gas was distilled by washing and compressing it through small stones in water. Pain used the gas for cooking and producing electricity. He also fueled a light van. Pain estimated that 10 kilograms (22 pounds) of brushwood would supply the gas equivalent of one litre (0.22 imp gal; 0.26 US gal). of petrol.

It took about 90 days to produce 500 cubic metres (18,000 cu ft) of gas - enough to keep two ovens and three burner stoves going for a year. The methane-fueled combustion engine drove a generator that produced 100 watts of electricity. This charged a battery, providing the light needed. Skepticism has been leveled at Pain's estimates for methane extraction and it is not known if anyone has been able to reproduce his results.

Pain's compost heaps generated hot water via 200 metres (656 ft) of pipe buried inside the compost mound. The pipe was wrapped around the methane generator with an inlet for cold water and an outlet for hot water. The heat from the decomposing mass produced 4 litres per minute (0.88 imp gal/min; 1.1 US gal/min) of hot water heated to 60 degrees Celsius (140 degrees Fahrenheit) — enough to meet central heating, bathroom and kitchen requirements. The heap composted for nearly 18 months, after which it was dismantled. The humus was used to mulch soils.

Heater Silo

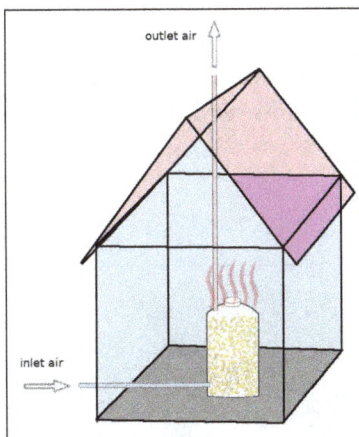

Using the heat of a compost heater silo for warming up the inside of a house.

The composting process runs in an airtight container inside a house. The heat can be radiated directly to the interiors of the house or distributed by a heating circuit. An additional water pipe can be vertically built into the silo for warming water.

The silo is the central part of an in-house compost heater. In autumn the silo is filled up with fresh biomass, after which the silo delivers comfortable heat throughout the winter.

Inlet air and outlet air provide the necessary oxygen. The outlet air goes out of the house. Silo moisture is higher than in a regular compost heap. The decay process produces additional water, which is drained at the bottom of the silo. A part of this water enters the top of the silo for better distribution in the processing volume. If pumped periodically or continuously to the top to rinse through the silo, the whole system becomes a wet composting system.

Rotational Grazing

Rotational grazing of cattle and sheep in Missouri with pasture divided into paddocks,
each grazed in turn for a period and then rested.

In agriculture, rotational grazing, as opposed to continuous grazing, describes many systems of pasturing, whereby livestock are moved to portions of the pasture, called paddocks, while the other portions rest. Each paddock must provide all the needs of the livestock, such as food, water and sometimes shade and shelter. The approach often produces lower outputs than more intensive animal farming operations, but requires lower inputs, and therefore sometimes produces higher net farm income per animal.

Approach

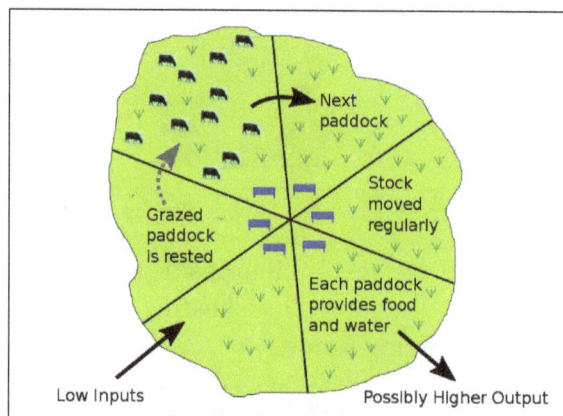

Diagram of rotational grazing, showing the use of paddocks, each providing food and water for the livestock for a chosen period.

In rotational grazing livestock are moved to portions of the pasture, called paddocks, while the other portions rest. The intent is to allow the pasture plants and soil time to recover.

Healing native rangeland may require a combination of burning and rotational grazing.

Rotational grazing can be used with ruminants such as beef or dairy cattle, sheep or goats, or even pigs. The herds graze one portion of pasture, or a paddock, while allowing the others to recover. The length of time a paddock is grazed will depend on the size of the herd and the size of the paddock and local environmental factors. Resting grazed lands allows the vegetation to regrow. Rotational grazing is especially effective because grazers do better on the more tender younger plant stems. These systems may or may not leave parasites behind to die off, minimizing or eliminating the need for de-wormers, depending if the rotational time is smaller or larger than the parasitic life cycle.

Benefits

Herd health benefits arise from animals having access to both space and fresh air. Freedom of movement within a paddock results in increased physical fitness, which limits the potential for injuries and abrasion, and sometimes depending on the system reduces the potential of exposure to high levels of harmful disease-causing microorganisms and insects.

In a concentrated animal feeding operation it is normal for a large number of animals to continuously occupy a small area. By comparison, with managed grazing, the animals are able to live in a more natural environment. The animals experience less disease and fewer foot ailments, depending on the rotational system being used.

Rotational grazing has been said to be more environmentally friendly in certain cases. Many pastures undergoing certain types of rotational grazing are less susceptible to soil erosion. Paddocks might require fewer inputs. These grazing regimes are sometimes said to be more resilient and more capable of responding to changing environmental conditions.

Problems

A key element of this style of animal husbandry is that either each grazed area must contain all elements needed for the animals (water source, for instance) or the feed or water source must be moved each time the animals are moved. Having fixed feeding or watering stations can defeat the rotational aspect, leading to degradation of the ground around the water supply or feed supply if additional feed is provided to the animals. Special care must be taken to ensure that high use areas do not become areas where mud, parasites or diseases are spread or communicated.

Several problems are related to shade in pasture areas. Although shade provides relief from heat and reduces the risk of heat stress, animals tend to congregate in these areas which leads to nutrient loading, uneven grazing, and potential soil erosion.

Ruminal tympany, also known as bloat, is a common serious problem when grazing ruminants on fresh, young pasture, and if left untreated can be fatal. This problem occurs when foam producing compounds in plants are digested by cows, causing foam to form in the rumen of the animal and not allowing animals to properly belch gas. Animals are especially susceptible to bloat if they are moved to new pasture sources when they are particularly hungry and especially on young, fresh and wet legumes. It is therefore important to ensure that the herd is eating enough at the end of a rotation when forage will be more scarce, limiting the potential for animals to gorge themselves when turned out onto new paddocks. The risk of bloat can be mitigated by careful management of rotations, seeding the non-bloating European legume species Lotus corniculatus in pasturelands, reducing the amount of legumes/increasing grasses, providing sufficient supplemental feeding and extra fodder when turning out on new paddocks, reducing the size of the paddock when livestock is first turned out, and daily rations of the anti-foaming agent poloxalene mixed well into the fodder.

Weed Control

A well managed rotational grazing system has low pasture weed establishment because the majority of niches are already filled with established forage species, making it harder for weeds to compete and become established. The use of multiple species in the pasture helps to minimize weeds. Established forage plants in rotational grazing pasture systems are healthy and unstressed due to the "rest" period, enhancing the competitive advantage of the forage. Additionally, in comparison to grain crop production, many plants which would be considered weeds are not problematic in perennial pasture. However, certain species such as thistles and various other weeds, are indigestible or poisonous to grazers. These plant species will not be grazed by the herd and can be recognized for their prevalence in pasture systems.

A key step in managing weeds in any pasture system is identification. Once the undesired species in a pasture system are identified, an integrated approach of management can be implemented to control weed populations. It is important to recognize that no single approach to weed management will result in weed free pastures; therefore, various cultural, mechanical, and chemical control methods can be combined in an weed management plan. Cultural controls include: avoiding spreading manure contaminated with weed seeds, cleaning equipment after working in weed infested areas, and managing weed problems in fencerows and other areas near pastures. Mechanical controls such as repeated mowing, clipping, and hand weeding can also be used to effectively manage weed infestations by weakening the plant. These methods should be implemented when weed flower buds are closed or just starting to open to prevent seed production. Although these first two methods reduce need for herbicides, weed problems may still persist in managed grazing systems and the use of herbicides may become necessary. Use of herbicides may restrict the use of a pasture for some length of time, depending on the type and amount of the chemical used. Frequently, weeds in pasture systems are patchy and therefore spot treatment of herbicides may be used as a least cost method of chemical control.

Nutrient Availability and Soil Fertility

If pasture systems are seeded with more than 40% legumes, commercial nitrogen fertilization is unnecessary for adequate plant growth. Legumes are able to fix atmospheric nitrogen, thus providing nitrogen for themselves and surrounding plants. Although grazers remove nutrient sources from the pasture system when they feed on forage sources, the majority of the nutrients consumed by the herd are returned to the pasture system through manure. At a relatively high stocking rate, or high ratio of animals per hectare, manure will be evenly distributed across the pasture system. The nutrient content in these manure sources should be adequate to meet plant requirements, making commercial fertilization unnecessary. Rotational grazing systems are often associated with increased soil fertility which arises because manure is a rich source of organic matter that increases the health of soil. In addition, these pasture system are less susceptible to erosion because the land base has continuous ground cover throughout the year.

High levels of fertilizers entering waterways are a pertinent environmental concern associated with agricultural systems. However, rotational grazing systems effectively reduce the amount of nutrients that move off-farm which have the potential to cause environmental degradation. These systems are fertilized with on-farm sources, and are less prone to leaching as compared to commercial fertilizers. Additionally, the system is less prone to excess nutrient fertilization, so the majority of nutrients put into the system by manure sources are utilized for plant growth. Permanent pasture systems also have deeper, better established forage root systems which are more efficient at taking up nutrients from within the soil profile.

Economics

Although milk yields are often lower in rotational systems, net farm income per cow is often greater as compared to confinement operations. This is due to the additional costs associated with herd health and purchased feeds are greatly reduced in management intensive rotational grazing systems. Additionally, a transition to rotational grazing is associated with low start-up and maintenance costs. Another consideration is that while production per cow is less, the number of cows per acre on the pasture can increase. The net effect is more productivity per acre at less cost.

The main costs associated with transitioning rotational grazing are purchasing fencing, fencers, and water supply materials. If a pasture was continuously grazed in the past, likely capital has already been invested in fencing and a fencer system. Cost savings to graziers can also be recognized when one considers that many of the costs associated with livestock operations are transmitted to the grazers. For example, the grazers actively harvest their own sources of food for the portion of the year where grazing is possible. This translates into lower costs for feed production and harvesting, which are fuel intensive endeavors. Rotational grazing systems rely on the grazers to produce fertilizer sources via their excretion. There is also no need for collection, storage, transportation, and application of manure, which are also all fuel intensive. Additionally, external fertilizer use contributes to other costs such as labor, purchasing costs.

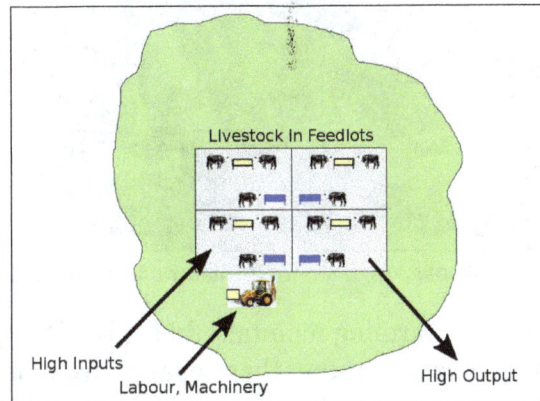

Confinement operations are more intensive, producing higher outputs for that land but requiring higher inputs from other acreage and additional labour and machinery, so rotational grazing often generates greater net farm income per cow.

Rotational grazing results in time savings because the majority of work which might otherwise require human labor is transmitted to the herd.

Natural Farming

Natural farming is an ecological farming approach established by Masanobu Fukuoka (1913–2008), a Japanese farmer and philosopher, introduced in his 1975 book The One-Straw Revolution.

The system works along with the natural biodiversity of each farmed area, encouraging the complexity of living organisms—both plant and animal—that shape each particular ecosystem to thrive along with food plants. Fukuoka saw farming both as a means of producing food and as an aesthetic or spiritual approach to life, the ultimate goal of which was, "the cultivation and perfection of human beings". He suggested that farmers could benefit from closely observing local conditions. Natural farming is a closed system, one that demands no human-supplied inputs and mimics nature.

Fukuoka's ideas radically challenged conventions that are core to modern agro-industries; instead of promoting importation of nutrients and chemicals, he suggested an approach that takes advantage of the local environment. Although natural farming is considered a subset of organic farming, it differs greatly from conventional organic farming, which Fukuoka considered to be another modern technique that disturbs nature.

Fukuoka claimed that his approach prevents water pollution, biodiversity loss and soil erosion, while providing ample amounts of food.

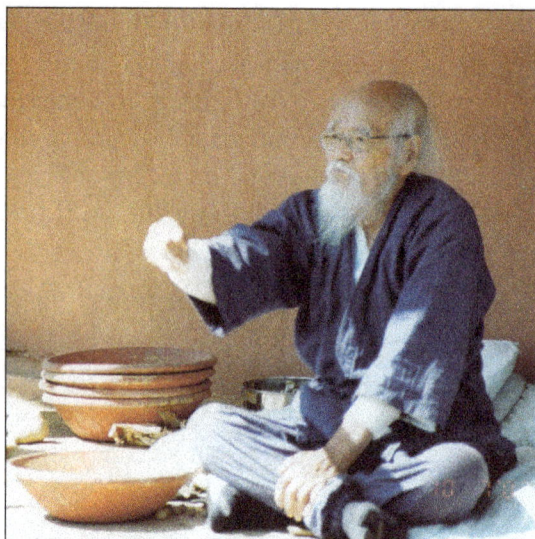

Masanobu Fukuoka, originator of the natural farming method.

In principal, practitioners of natural farming maintain that it is not a technique but a view, or a way of seeing ourselves as a part of nature, rather than separate from or above it. Accordingly, the methods themselves vary widely depending on culture and local conditions.

Rather than offering a structured method, Fukuoka distilled the natural farming mindset into five principles:

- No tillage,

- No fertilizer,

- No pesticides or herbicides,

- No weeding,

- No pruning.

A young man helps harvest rice by hand at a natural farm, in this production still from the film "Final Straw: Food, Earth, Happiness".

Though many of his plant varieties and practices relate specifically to Japan and even to local conditions in subtropical western Shikoku, his philosophy and the governing principles of his farming systems have been applied widely around the world, from Africa to the temperate northern hemisphere.

Principally, natural farming minimises human labour and adopts, as closely as practical, nature's production of foods such as rice, barley, daikon or citrus in biodiverse agricultural ecosystems. Without plowing, seeds germinate well on the surface if site conditions meet the needs of the seeds placed there. Fukuoka used the presence of spiders in his fields as a key performance indicator of sustainability.

Fukuoka specifies that the ground remain covered by weeds, white clover, alfalfa, herbaceous legumes, and sometimes deliberately sown herbaceous plants. Ground cover is present along with grain, vegetable crops and orchards. Chickens run free in orchards and ducks and carp populate rice fields.

Periodically ground layer plants including weeds may be cut and left on the surface, returning their nutrients to the soil, while suppressing weed growth. This also facilitates the sowing of seeds in the same area because the dense ground layer hides the seeds from animals such as birds.

For summer rice and winter barley grain crops, ground cover enhances nitrogen fixation. Straw from the previous crop mulches the topsoil. Each grain crop is sown before the previous one is harvested by broadcasting the seed among the standing crop. Later, this method was reduced to a single direct seeding of clover, barley and rice over the standing heads of rice. The result is a denser crop of smaller, but highly productive and stronger plants.

Fukuoka's practice and philosophy emphasised small scale operation and challenged the need for mechanised farming techniques for high productivity, efficiency and economies of scale. While his family's farm was larger than the Japanese average, he used one field of grain crops as a small-scale example of his system.

Climax Ecosystems

In ecology, climax ecosystems are mature ecosystems that have reached a high degree of stability, productivity and diversity. Natural farmers attempt to mimic those virtues, creating a comparable climax ecosystem, and employ advanced techniques such as intercropping, companion planting and integrated pest management.

No-till

Natural farming recognizes soils as a fundamental natural asset. Ancient soils possess physical and chemical attributes that render them capable of generating and supporting life abundance. It can be argued that tilling actually degrades the delicate balance of a climax soil:

- Tilling may destroy crucial physical characteristics of a soil such as water suction, its ability to send moisture upwards, even during dry spells. The effect is due to pressure differences between soil areas. Furthermore, tilling most certainly destroys soil horizons and hence disrupts the established flow of nutrients. A study suggests that reduced tillage preserves the crop residues on the top of the soil, allowing organic matter to be formed more easily and hence increasing the total organic carbon and nitrogen when compared to conventional tillage. The increases in organic carbon and nitrogen increase aerobic, facultative anaerobic and anaerobic bacteria populations.

- Tilling over-pumps oxygen to local soil residents, such as bacteria and fungi. As a result, the chemistry of the soil changes. Biological decomposition accelerates and the microbiota mass increases at the expense of other organic matter, adversely affecting most plants, including trees and vegetables. For plants to thrive a certain quantity of organic matter (around 5%) must be present in the soil.

- Tilling uproots all the plants in the area, turning their roots into food for bacteria and fungi. This damages their ability to aerate the soil. Living roots drill millions of tiny holes in the soil and thus provide oxygen. They also create room for beneficial insects and annelids (the phylum of worms). Some types of roots contribute directly to soil fertility by funding a mutualistic relationship with certain kinds of bacteria (most famously the rhizobium) that can fix nitrogen.

Fukuoka advocated avoiding any change in the natural landscape. This idea differs significantly from some recent permaculture practice that focuses on permaculture design, which may involve the change in landscape. For example, Sepp Holzer, an Austrian permaculture farmer, advocates the creation of terraces on slopes to control soil erosion. Fukuoka avoided the creation of terraces in his farm, even though terraces were common in China and Japan in his time. Instead, he prevented soil erosion by simply growing trees and shrubs on slopes.

Synergistic Gardening

It's a method of farming that respects Nature and at the same time can guarantee healthy and abundant crops. Synergistic Agriculture promotes a way of farming that uses the self-fertility of wild soil as fertilizer and its principles can be applied both in wide and small spaces, city or countryside, in cold and warm climates.

Fundamental Principles

Synergistic Agriculture respects the following 4 principles:

- Do not work the soil: No ploughing or hoeing. The soil shouldn't be disturbed and by turning the soil over we interrupt the combined action of plant roots and chemical activity of bacteria, fungi and worms.

- Do not fertilize: The soil is fertilized by a permanent organic cover and the roots of plants. Roots, in particular, should never be pulled up, they remain in the soil and the leaves are left where they fall.

- Do not use chemical pests or fertilizers: The use of such substances is neither natural nor sustainable.

- Do not compact the soil: In order for the micro-ecosystems found in the soil to be sufficiently aerated, it is necessary to avoid compacting the soil.

Synergistic Agriculture in Practice

First of all it necessary to prepare the land. This step should not be taken for granted and time and procedures depend on the initial condition of the ground (i.e. a field previously cultivated organically; a ploughed land treated with chemical products, etc.). Regardless of the kind of ground, you'll have to make sure that there is no plough pan, that is to say a compacted layer in cultivated soil resulting from repeated ploughing. In order to determine whether it's present or not, we suggest to dig a hole (50 cm deep) and check if there is a layer of different colour than the rest of the ground. In case you find it, you will have to work the land with a ripper or by hand with a spade fork in order to break this layer of mineral salts. The second step is to build a wind fence that will also be a great shelter for good insects and predators of parasites, as well as for the hibernation of hedgehogs and bird nests.

The next stage is the preparation of the raised beds where plants are cultivated. The beds can be straight, curve or mandala-shaped and their function is to mark the passages where people walk and work in order to avoid compacting the cultivated land. We suggest the beds to be 120 cm wide, while there is no limit of length. It is advisable to create passages 80 cm wide at least every 4-5 metres. The maximum height of the beds should be 30-40 cm, otherwise the cultivable surface would be too much reduced. The beds must be built with the same land of the vegetable garden you intend to create and a drip irrigation must be installed on them.

If the quality of your soil is good, there is no need to mix compost to the soil while building the beds, otherwise you can incorporate some superficially. The beds will have two cultivable surfaces:

- The central area, where the ground is deeper;

- The sides, whose height ranges between 15 and 40 cm.

When the beds are ready and before sowing any plant, some "permanent tutors" must be fixed to the ground. These tutors are arches that cross above the beds, they can be iron bars for building, diameter 12, as well as bamboo canes and their function is to guide climbing plants.

In the same bed at least 3 different families of vegetables must be planted in order to make the most of synergistic actions of plants. In general the sides of beds are used for cultivating plants that are complementary to the main crops. The plants to be sown in the sides are plants that grow vertically that belong to the lily family such as shallot, garlic, onion, leek, etc. This type of plants spread fungicide, antibacterial and insect repellent substances that will also protect your vegetables. Along with these plants you should also sow some fast-growing plants such as chicory or lettuce, leaving the roots in the ground when harvesting. On the edges, between the central area and the sides, a row of legumes, peas or beans should be planted depending on the season. It is important to always keep the sides covered by mulch because it tends to roll down toward the passages. Besides legumes, in the central area of the beds other vegetable families should be planted: crucifers, solanaceae, cucrbitaceae, apiaceae, lamiaceae, rosaceae, asteraceae, etc.

Importance of Mulching

Mulch is a very important and a not-to-be-missed element of Synergistic Agriculture because it prevents the soil to compact and it protects from the consequences excessive sunlight, rain and wind. Moreover, it helps the development of microorganisms and especially earthworms. Mulch acts like a thermal cover, protecting the soil both in warm and cold seasons. Obviously mulch has to be a biodegradable material depending on our needs, such as straws, canes, leaves, paper, cardboard, etc.

Ultimate Aim of Synergistic Agriculture

Nowadays we depend on agriculture for feeding a constantly-growing world population. However, agriculture needs to take into account the life of the lands and the environment. Emilia's aim was reproducing in agriculture the same soil conditions that in nature produce such perfect crops and, at the same time, building a new future where the respect for life is applied through our synergistic and environmental integration.

References

- Iverson, Aaron L.; Marín, Linda E.; Ennis, Katherine K.; Gonthier, David J.; Connor-Barrie, Benjamin T.; Remfert, Jane L.; Cardinale, Bradley J.; Perfecto, Ivette (2014-10-03). "REVIEW: Do polycultures promote win-wins or trade-offs in agricultural ecosystem services? A meta-analysis". Journal of Applied Ecology. 51 (6): 1593–1602. doi:10.1111/1365-2664.12334

- Ferguson, Bruce; et al. (18 June 2013). "Sustainability of holistic and conventional cattle ranching in the seasonally dry tropics of Chiapas, Mexico". Agricultural Systems. 120: 38–48. doi:10.1016/j.agsy.2013.05.005

- Falk, Ben (2013). The Resilient Farm and Homestead: An Innovative Permaculture and Whole Systems Design Approach. Chelsea Green Publishing. p. 280. ISBN 978-1-60358-444-9

- Randolph, R., & McKay, C. (2014). Protecting and expanding the richness and diversity of life, an ethic for astrobiology research and space exploration. International Journal of Astrobiology, 13(1), 28-34. doi:10.1017/S1473550413000311

- Priya Reddy; Prescott College Environmental studies (2010). Sustainable Agricultural Education: An Experiential Approach to Shifting Consciousness and Practices. Prescott College. ISBN 978-1-124-38302-6

- Synergistic-agriculture: autosufficienza.it, Retrieved 27 June, 2019

Common Practices in Permaculture

3

- **Natural Building**
- **Agroforestry**
- **Hügelkultur**
- **Fruit Tree Pruning**
- **Intensive Rotational Grazing**

There is a diverse range of practices that are followed in permaculture. Natural building, agroforestry, hügelkultur, fruit tree pruning and management, intensive rotational grazing, etc. are some practices that fall in its domain. The topics elaborated in this chapter will help in gaining a better perspective about different practices in permaculture.

Natural Building

Natural building gives you the chance to build with a much lower energy footprint through using available, more local materials – rather than things like cement, clay brick and treated timber that have huge flow-on effects to both the environments they're sourced from, and the earth's atmosphere because of how they're produced.

But like designing and building anything, your context – which includes your site, climate, resources, location and budget – will define what methods you use to build.

Exterior and Interior Walls

The first consideration is that not all walls are equal – your exterior and interior walls perform very different functions. We love this quote from Sam Vivas, the master builder who teaches our Natural Building courses, talking about walls of a house:

> "A livable house should have a good blanket around the outside, and a hot water bottle on the inside".

The 'good blanket' is the exterior walls, meaning that they should be as insulative as possible, to protect the inside of the house from outdoor temperature fluctuations of both hot and cold.

The 'hot water bottle' is the interior walls, and floor if you can manage it, meaning that they should have as much thermal mass as possible.

These thermal mass banks then become an important passive heat sink for the living space, stabilising the inside temperature of the house with a minimum of extra heating or cooling, and making the house super comfortable in both summer and winter. Different methods in natural building are:

Strawbale Walls – infill + load bearing

Strawbales are a favourite exterior wall choice for many natural builders, with good reason. They can be used to build strong walls with a very high insulation value, they're highly fire resistant when constructed right, and they go up quickly. Bonuses include that the straw is a waste product of agriculture, and it's possible to source reasonably local material anywhere.

As a product of industrialised agriculture, strawbales come in standard sizes, which allows you to plan ahead when using them. There's two main ways that they're used for exterior walls – as a load-bearing arrangement (with no other uprights, just the roof sitting on top) or as infill, with a wooden or steel frame creating the main frame of the structure.

- Load-bearing strawbale walls – this technique sees the walls of a structure go up all at once, and is less complex in some ways, as you're not working around wall frames, so it can be quicker that infill. However, load bearing walls need to all go up at once before any more building can occur, and of course, if it rains before you've completed the walls and got a roof on it's a BIG problem because your bales might get wet.

- In-fill strawbale walls – this technique is the main way strawbale buildings are constructed, in Australia at least. Far less time-critical and heart stopping than building load bearing strawbale walls, infill sees a building framed up first, the main roof goes on, and only then do you start putting in the exterior wall strawbales. Much less panic inducing if clouds are on the horizon.

The infill technique has the great advantage of giving you a big roof to work under (a much loved aspect of any building site), and also allows other internal building works to proceed while the exterior walls are filled with strawbales. Infill strawbale buildings are also typically easier to get approval for from councils, as the technique is seen as less 'weird' than load-bearing strawbale.

Rammed Earth Walls

Rammed earth walls are a popular choice for interior walls because of their high thermal mass. Made up of earth, sand and a bit of cement, the majority of their components can be sourced locally in most places.

Because the earth is literally and mechanically 'rammed' into formwork to make rammed earth walls, they do need the building to have very strong foundations in order to be built. And if a second level is going on, the floor level internal rammed earth walls need to be quite wide to bear the weight above, which can lead to large wall footprints in your building.

Light Earth Walls

Light earth is a funky blend between rammed earth and cobb. It comprises of loose straw coated in a small amount of mud, then packed into formwork to make walls, with the formwork being removed once the wall is semi-dry.

Light earth is a good choice for interior walls (not load-bearing ones though, this is an in-fill method) for home builders because it's an easy technique, cheap to make, extremely effective, and has a higher thermal mass than strawbale (though not as high as rammed earth).

What you save in material costs with light earth you spend in human energy – so while it's a lot less labour intensive than a technique like cobb, there's till some hands-on involved. Whack, whack, whack. Make sure you have lots of friends to help, and good food and drinks waiting after the whacking sessions.

Cobb Walls

Cobb walls are an old, old, old wall building technique hailing from many continents. A mix of mud, loose straw and sand are combined in a vessel, and then handfuls are applied to slowly build a wall, a bit like a sand castle.

This slowly creates a strong wall with great thermal mass. Like light earth, cobb wall building is simple, cheap on materials, and relies on human energy rather than brought-in solutions to make and create the walls.

There's other wall methods:

- Cordwood – a mix between cobb and wood – this technique has a high thermal mass and looks gorgeous.

- Super adobe building.

- Earthship style wall construction.

- Constructed straw panels (think fibre board, but made of straw) which are an off-the-shelf solution but can be perfect for some scenarios, like retrofitting exterior mudbrick walls to provide some exterior insulation.

A natural building is an opportunity to see the land, resources and environment around you in a different way, and to find ways to make them all a part of your home.

Agroforestry

Agroforestry is a land use management system in which trees or shrubs are grown around or among crops or pastureland. This intentional combination of agriculture and forestry has varied benefits, including increased biodiversity and reduced erosion. Agroforestry practices have been successful in sub-Saharan Africa and in parts of the United States.

Agroforestry shares principles with intercropping. Both may place two or more plant species (such as nitrogen-fixing plants) in close proximity.

Agroforestry in Burkina Faso: maize grown under Faidherbia albida and Borassus akeassii near Banfora.

Agroforestry contour planting integrated with animal grazing on Taylors Run farm, Australia.

As a Science

According to Wojtkowski, the theoretical base for agroforestry lies in ecology, or agroecology. Agroecology encompasses diverse applications such as countering winds, high rainfall, harmful insects, etc. From this perspective, agroforestry is one of the three principal agricultural land-use sciences. The other two are agriculture and forestry.

Benefits

Agroforestry systems can be advantageous over conventional agricultural and forest production methods. They can offer increased productivity, economic benefits, and more diversity in the ecological goods and services provided.

Biodiversity

Biodiversity in agroforestry systems is typically higher than in conventional agricultural systems. Two or more interacting plant species in a given area create a more complex habitat that can support a wider variety of fauna.

Agroforestry is important for biodiversity for different reasons. It provides a more diverse habitat than a conventional agricultural system. Tropical bat and bird diversity for instance can be comparable to the diversity in natural forests. Although agroforestry systems do not provide as many floristic species as forests and do not show the same canopy height, they do provide food and nesting possibilities. A further contribution to biodiversity is that the germplasm of sensitive species can be preserved. As agroforests have no natural clear areas, habitats are more uniform. Furthermore, agroforests can serve as corridors between habitats. Agroforestry can help to conserve biodiversity by having a positive influence on other ecosystem services.

Soil and Plant Growth

Depleted soils can be protected from soil erosion by groundcover plants such as naturally growing grasses in agroforestry systems. These help to stabilise the soil as they increase cover compared to short-cycle cropping systems. Soil cover is a crucial factor in preventing erosion. Cleaner water through reduced nutrient and soil surface runoff can be a further advantage of agroforestry. The runoff can be reduced by decreasing its velocity and increasing infiltration into the soil. Compared to row-cropped fields nutrient uptake can be higher and reduce nutrient loss into streams.

Further advantages concerning plant growth:

- Bioremediation.

- Drought resistance.

- Increased crop stability.

Contribution to Sustainable Agricultural Systems

- Reduced poverty through increased production of wood and other products.

- Increased food security by restored soil fertility for food crops.

- Multifunctional site use, e.g., crop production and animal grazing.

- Reduced global warming and hunger risk by increasing the number of drought-resistant trees and the subsequent production of fruits, nuts and edible oils.

- Reduced deforestation and pressure on woodlands by providing farm-grown fuelwood.

- Reduced need for toxic chemicals (insecticides, herbicides, etc.).

- Improved human nutrition through more diverse farm outputs.

- Growing space for medicinal plants e.g., in situations where people have limited access to mainstream medicines.

Other Environmental Goals

Carbon sequestration is an important ecosystem service. Trees in agroforestry systems, like in new forests, can recapture some of the carbon that was lost by cutting existing forests. They also provide additional food and products. The rotation age and the use of the resulting products are important factors controlling the amount of carbon sequestered. Agroforests can reduce pressure on primary forests by providing forest products.

Agroforestry practices may realize a number of environmental goals, such as:

- Odour, dust and noise reduction.

- Green space and visual aesthetics.

- Enhancement or maintenance of wildlife habitat.

Adaptation to Climate Change

Especially in recent years, poor smallholder farmers turned to agroforestry as a mean to adapt to climate change. A study from the CGIAR research program on Climate Change, Agriculture and Food Security (CCAFS) found from a survey of over 700 households in East Africa that at least 50% of those households had begun planting trees in a change from earlier practices. The trees were planted with fruit, tea, coffee, oil, fodder and medicinal products in addition to their usual harvest. Agroforestry was one of the most widespread adaptation strategies, along with the use of improved crop varieties and intercropping.

Applications

Tropical Agroforestry

Research with Faidherbia albida in Zambia showed maximum maize yields of 4.0 tonnes per hectare using fertilizer and inter-cropped with these trees at densities of 25 to 100 trees per hectare, compared to average maize yields in Zimbabwe of 1.1 tonnes per hectare.

Hillside Systems

A well-studied example of an agroforestry hillside system is the Quesungual Slash and Mulch Agroforestry System (QSMAS) in Lempira Department, Honduras. This region was historically used for slash and burn subsistence agriculture. Due to heavy seasonal floods, the exposed soil was washed away, leaving infertile barren soil exposed to the dry season. Farmed hillside sites had to be abandoned after a few years and new forest was burned. The Food and Agriculture Organization of the United Nations (FAO) helped introduce a system incorporating local knowledge consisting of the following steps:

- Thin and prune Hillside secondary forest, leaving individual beneficial trees, especially nitrogen-fixing trees. They help reduce soil erosion, maintain soil moisture, provide shade and provide an input of nitrogen-rich organic matter in the form of litter.

- Plant maize in rows. This is a traditional local crop.

- Harvest from the dried plant and plant beans. The maize stalks provide an ideal structure for the climbing bean plants. Bean is a nitrogen-fixing plant and therefore helps introduce more nitrogen.

- Pumpkin can be planted during this time. Its large leaves and horizontal growth provide additional shade and moisture retention. It does not compete with the beans for sunlight since the latter grow vertically on the stalks.

- Every few seasons, rotate the crop by grazing cattle, allowing grass to grow and adding soil organic matter and nutrients (manure). The cattle prevent total reforestation by grazing around the trees.

- Repeat.

Shade Crops

With shade applications, crops are purposely raised under tree canopies within the shady environment. The understory crops are shade tolerant or the overstory trees have fairly open canopies. A conspicuous example is shade-grown coffee. This practice reduces weeding costs and improves coffee quality and taste.

Crop-over-tree Systems

Crop-over-tree systems employ woody perennials in the role of a cover crop. For this, small shrubs or trees pruned to near ground level are utilized. The purpose is to increase in-soil nutrients and to reduce soil erosion.

Intercropping and Alley Cropping

With alley cropping, crop strips alternate with rows of closely spaced tree or hedge species. Normally, the trees are pruned before planting the crop. The cut leafy material is spread over the crop area to provide nutrients. In addition to nutrients, the hedges serve as windbreaks and reduce erosion.

In tropical areas of North and South America, various species of Inga such as I. edulis and I. oerstediana have been used for alley cropping.

Intercropping is advantageous in Africa, particularly in relation to improving maize yields in the sub-Saharan region. Use relies upon the nitrogen-fixing tree species Sesbania sesban, Tephrosia vogelii, Gliricidia sepium and Faidherbia albida. In one example, a ten-year experiment in Malawi showed that, by using the fertilizer tree Gliricidia (Gliricidia sepium) on land on which no mineral fertilizer was applied, maize yields averaged 3.3 tonnes per hectare as compared to one tonne per hectare in plots without fertilizer trees or mineral fertilizers.

Taungya

Taungya is a system originating in Burma. In the initial stages of an orchard or tree plantation, trees are small and widely spaced. The free space between the newly planted trees accommodates a seasonal crop. Instead of costly weeding, the underutilized area provides an additional output and income. More complex taungyas use between-tree space for multiple crops. The crops become more shade tolerant as the tree canopies grow and the amount of sunlight reaching the ground declines. Thinning can maintain sunlight levels.

Temperate Agroforestry

Although originally a concept used in tropical agronomy, the USDA distinguishes five applications of agroforestry for temperate climates.

Alley Cropping and Strip Cropping

Alley cropping corn fields between rows of walnut trees.

Alley cropping can also be used in temperate climates. Strip cropping is similar to alley cropping in that trees alternate with crops. The difference is that, with alley cropping, the trees are in single row. With strip cropping, the trees or shrubs are planted in wide strip. The purpose can be, as with alley cropping, to provide nutrients, in leaf form, to the crop. With strip cropping, the trees can have a purely productive role, providing fruits, nuts, etc. while, at the same time, protecting nearby crops from soil erosion and harmful winds.

Fauna-based Systems

Silvopasture over the years (Australia).

Trees can benefit fauna. The most common examples are silvopasture where cattle, goats, or sheep browse on grasses grown under trees. In hot climates, the animals are less stressed and put on weight faster when grazing in a cooler, shaded environment. The leaves of trees or shrubs can also serve as fodder.

Similar systems support other fauna. Deer and hogs gain when living and feeding in a forest ecosystem, especially when the tree forage nourishes them. In aquaforestry, trees shade fish ponds. In many cases, the fish eat the leaves or fruit from the trees.

The dehesa or montado system of silviculture are an example of pigs and bulls being held extensively in Spain and Portugal.

Boundary Systems

- A living fence can be a thick hedge or fence wire strung on living trees. In addition to restricting the movement of people and animals, living fences offer habitat to insect-eating birds and, in the case of a boundary hedge, slow soil erosion.

- Riparian buffers are strips of permanent vegetation located along or near active watercourses or in ditches where water runoff concentrates. The purpose is to keep nutrients and soil from contaminating the water.

- Windbreaks reduce wind velocity over and around crops. This increases yields through reduced drying of the crop and by preventing the crop from toppling in strong wind gusts.

A riparian buffer bordering a river in Iowa.

Challenges

Although agroforestry systems can be advantageous, they are not widespread in the US as of 2013.

As suggested by a survey of extension programs in the United States, obstacles (ordered most critical to least critical) to agroforestry adoption include:

- Lack of developed markets.

- Unfamiliarity with technologies.

- Lack of awareness.

- Competition between trees, crops and animals.

- Lack of financial assistance.

- Lack of apparent profit potential.

- Lack of demonstration sites.

- Expense of additional management.

- Lack of training or expertise.

- Lack of knowledge about where to market products.

- Lack of technical assistance.

- Adoption/start up costs, including costs of time.

- Unfamiliarity with alternative marketing approaches (e.g. web).

- Unavailability of information about agroforestry.

- Apparent inconvenience.

- Lack of infrastructure (e.g. buildings, equipment).

- Lack of equipment.

- Insufficient land.

- Lack of seed/seedling sources.

- Lack of scientific research.

Some solutions to these obstacles have been suggested.

Hügelkultur

Hügelkultur is a horticultural technique where a mound constructed from decaying wood debris and other compostable biomass plant materials is later (or immediately) planted as a raised bed. Adopted by permaculture advocates, it is suggested the technique helps to improve soil fertility, water retention, and soil warming, thus benefiting plants grown on or near such mounds.

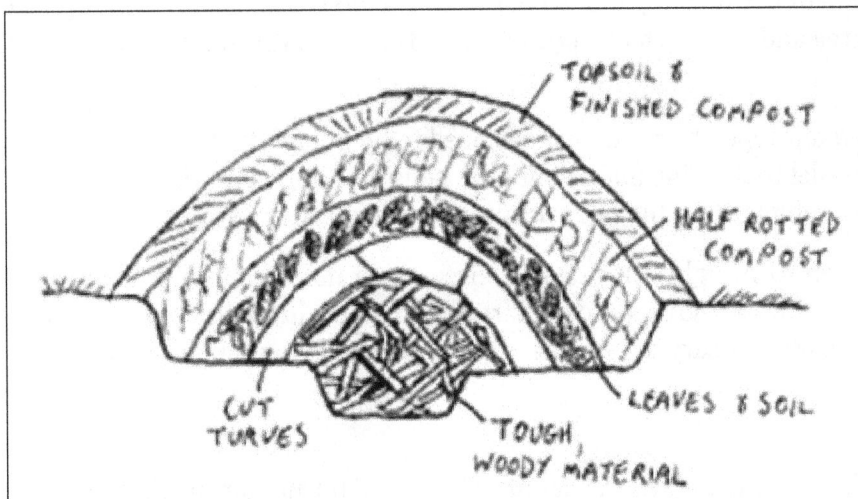

Layers of woody material and compost or soil used to build a Hugelkultur bed.

Use

Construction

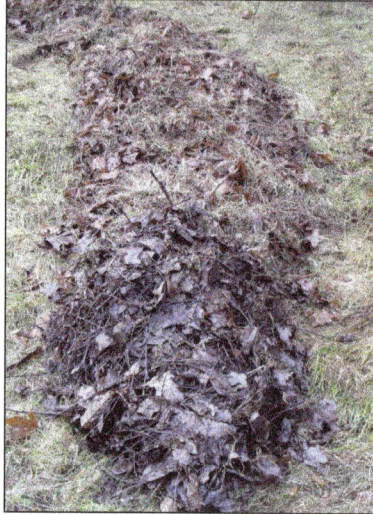

Hügelkultur bed construction, shown without the top layer of soil.

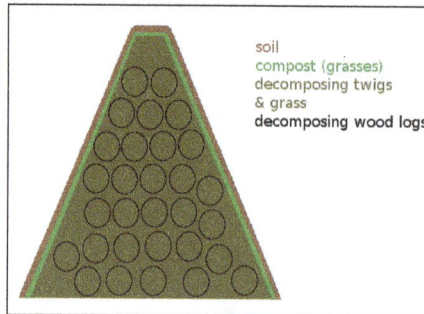

Hügelkultur layout.

In its basic form, mounds are constructed by piling logs, branches, plant waste, compost and additional soil directly on the ground. The pile has the form of a pyramid. The sides of the two slopes both have a grade of between 65 and 80 degrees. The beds are usually about 3 feet (0.91 m) by 6 feet (1.8 m) in area and about 3 feet (0.91 m) high. However, this height reduces as decomposition progresses.

When positioned on sloped terrain, the beds need to be put at an angle to the hillside (rather than having them parallel to it). This makes sure the beds do not receive unequal amounts of water. In most cases, it is useful to have the beds positioned against the prevailing wind direction.

The raised bed can form light-duty swales, circles and mazes. Mounds may also be made from alternating layers of wood, sod, compost, straw, and soil. Although their construction is straightforward, planning is necessary to prevent steep slopes that would result in erosion.

Planting

The mound is left to rest for several months before planting, although some advise immediate planting.

Anything can be grown on the raised beds, but if the bed will decompose/release its nutrients quickly (so long as it is not made of bulky materials like tree trunks), more demanding crops such as pumpkins, courgettes, cucumbers, cabbages, tomatoes, sweet corn, celery, or potatoes are grown in the first year, after which the bed is used for less demanding crops like beans, peas, and strawberries.

Hügelkultur is said to replicate the natural process of decomposition that occurs on forest floors, however in natural ecosystems wood would be present at the soil surface. Trees that fall in a forest often become nurse logs decaying and providing ecological facilitation to seedlings. As the wood decays, its porosity increases, allowing it to store water like a sponge. The water is slowly released back into the environment, benefiting nearby plants.

Hügelkultur beds are said to be ideal for areas where the underlying soil is of poor quality or compacted. They tend to be easier to maintain due to their relative height above the ground.

Decomposition speed of organic material depends on the carbon to nitrogen ratio of the material, among other factors. Wood breaks down relatively slowly because it has one of the highest carbon to nitrogen ratios of all organic matter that is used in composting. If the wood is not processed into smaller pieces with larger surface area to speed up chemical reactions, breakdown is even slower. The decomposition process may in the short term take more nitrogen from the soil through microbial activity (nitrogen immobilization), if not enough nitrogen is available. Thus in the short term the fertility of the soil may be decreased before eventually, perhaps after 1–2 years, the nitrogen level is increased past the original level. Traditionally therefore, it is said to be advantageous to balance "browns" (e.g. woodchippings) with "greens" (e.g. leaves) for efficient composting, and to allow compost to become well-rotted before applying it a bed to prevent competition between soil bacteria and plants for nitrogen, reducing yield.

Hügelkultur Mounds as Solid Earthworks

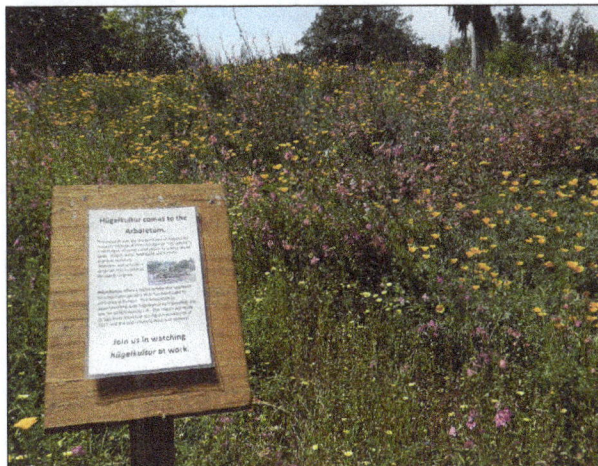

Hügelkultur bed with wildflower overplanting.

Although Hügelkultur beds can safely retain water in light-duty applications (for example, conserving the moisture of rain that falls on the bed), creating heavy-duty rainwater retention areas behind Hügelkultur beds on contour, to catch surface runoff from surrounding areas, can be dangerous. Some designers conflate the Hügelkultur bed's appearance with that of solid earthworks, but Hügelkultur beds cannot predictably control large amounts of stormwater in the way that solid earthworks can. Whereas embankment dams or the hillsides of swales can be relied on to hold back many thousands of gallons of water for weeks to allow it to seep into the ground, and berms can slow runoff, Hügelkultur beds are different in two ways: earthworks have no buoyant core (whereas Hügelkultur mounds contain logs), and the soil that they are made of is compacted. If fresh or dried timber is used in the bed, it may become buoyant in the water-saturated substrate, bursting from the soil covering and releasing all the sitting water through a breach. This can be an issue for years, until the wood is sufficiently rotten and infused with water. Another consideration is that Hügelkultur beds will degrade, shrinking over time into much lower mounds of soft, rich soil. This means that the retention area will have less depth as time goes on, but it also means that the uncompacted soil will remain a threat to breaching even if the logs become saturated.

There is a recorded instance of a breach occurring in a new project. Upon the first rainstorm, the retention areas behind the Hügelkultur beds filled with water and broke through. The released water carried the freshly-buried logs and dirt downhill, smashing a hole in a building being used as a church and filling the space with mud. No injuries were reported.

Some permaculturists have taken mild positions against the "hügel swales" still being promoted by other permaculturists, citing the danger and cross-purposes of Hügelkultur beds and swales.

Overfertilization, Contamination of Soil and Water Habitats

Over-fertilized plants are said to have less flavor, and too much nitrogen can be consumed by eating certain plants which have been over-fertilised (e.g. spinach). Advocates state that overfertilization is a risk in the first year if woodchips are used, which will break down too fast. Instead raised beds made with whole logs release nutrients slowly over a period of years. It has been suggested that excessive use of decomposing organic matter in Hügelkultur could leach out and contaminate and disrupt soil and water habitats.

Fruit Tree Pruning

Fruit tree pruning is the cutting and removing of selected parts of a fruit tree. It spans a number of horticultural techniques. Pruning often means cutting branches back, sometimes removing smaller limbs entirely. It may also mean removal of young shoots, buds, and leaves.

Established orchard practice of both organic and nonorganic types typically includes pruning. Pruning can control growth, remove dead or diseased wood, and stimulate the formation of flowers and fruit buds. It is widely stated that careful attention to pruning and training young trees improves their later productivity and longevity, and that good pruning and training can also prevent later injury from weak crotches or forks (where a tree trunk splits into two or more branches) that break from the weight of fruit, snow, or ice on the branches.

Some sustainable agriculture or permaculture personalities, such as Sepp Holzer and Masanobu Fukuoka, advocate and practice no-pruning methods, which runs counter to the widespread confidence in the idea that pruning produces superior results compared with not pruning.

A community apple orchard originally planted for productive use during the 1920s, in Westcliff on Sea (Essex, England), illustrating long neglected trees that have recently been pruned to renew their health and cropping potential.

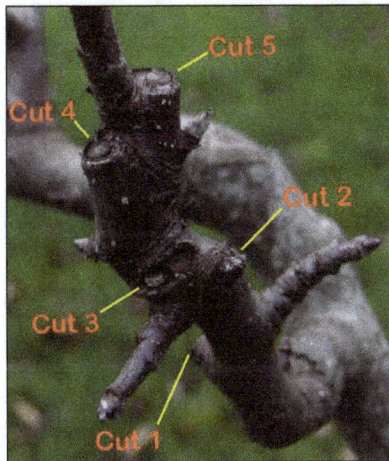

An apple tree sprout is being converted to a branched, fruit-bearing spur by an arborist.
Numbers show the sequence of cuts, which occurred during two years.

Plants form new tissue in an area called the meristem, located near the tips of roots and shoots, where active cell division takes place. Meristem growth is aimed at ensuring that leaves are quickly elevated into sunlight, and that roots are able to penetrate deeply into the soil. Once adequate height and length is achieved by the stems and roots, they begin to thicken to support the plant. On the shoots, these growing tips of the plant are called apical buds. The apical meristem (or tip) produces the growth hormone auxin, which not only promotes cell division, but also diffuses downwards and inhibits the development of lateral bud growth that otherwise competes with the apical tip for light and nutrients. Removing the apical tip and its suppressive hormone lets lower, dormant lateral buds develop, and the buds between the leaf stalk and stem produce new shoots that compete to become lead growth.

Manipulating this natural response to damage (known as the principle of apical dominance) by processes such as pruning (as well as coppicing and pollarding) allows the arborist to determine the shape, size, and productivity of many fruiting trees and bushes. The main aim when pruning fruit trees is usually to maximize fruit yield. Unpruned trees tend to produce large numbers of small fruits that may be difficult to reach when harvesting by hand. Branches can become broken by the weight of the crop, and the cropping may become biennial (that is, bearing fruit only every other year). Overpruned trees on the other hand tend to produce light crops of large, flavourless fruit that does not store well. Careful pruning balances shoot growth and fruit production.

Formative Pruning of Bush Trees

In the early years of the tree's life, it is important to develop a framework sufficiently strong to bear the weight of crops. This requires formative pruning to reinforce the tree. Formative pruning of apple (Malus pumila) and pear (Pyrus communis) trees should be carried out in the dormant winter months. For the Northern hemisphere, this should occur between November and March; For the Southern hemisphere, June and September. Stone fruits—such as cherries, plums, and gages—have different requirements, and should not be pruned in dormant months.

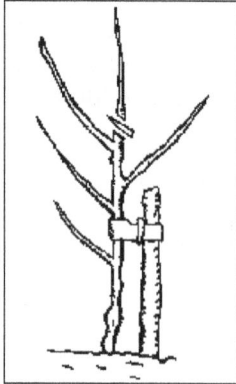
One-year-old tree.

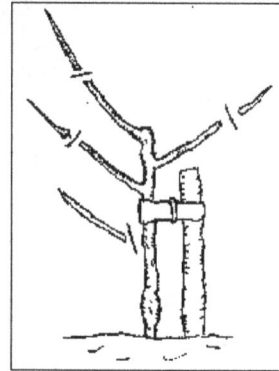
Two-year-old tree.

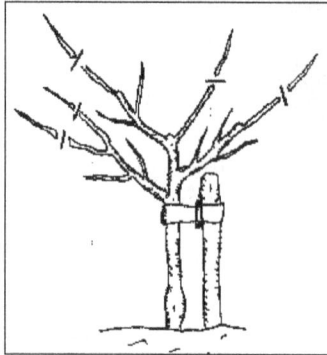
Three-year-old tree.

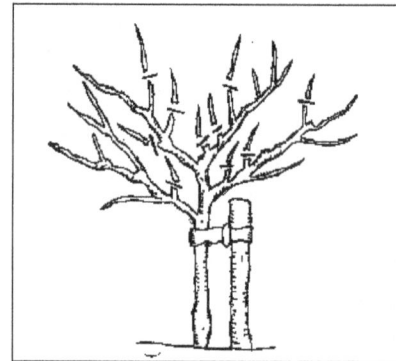
Four-year-old tree.

Maiden Tree

A maiden whip (a one-year-old tree with no side shoots) should be pruned to a bud with two buds below it at about 80 cm from the ground immediately after planting to produce primary branches during the first growing season. A feathered maiden (that is, a one-year-old tree with several side branches) should have its main stem pruned back to three or four strong shoots at 80 cm from the ground. Side shoots should be shortened by two thirds of their length to an upward or outward facing bud. Lower shoots should be removed flush with the stem.

Two Year

Remove lower shoots and prune between three and five of the best-placed shoots by half to an upwards or outwards facing bud to form what becomes the tree's main structural branches. Remove any inward-facing shoots.

Three Year

Prune leading shoots of branches selected to extend the framework by half, to a bud facing in the desired direction. Select four good laterals to fill the framework and shorten these by a half. Prune any remaining laterals to four buds to form fruiting spurs.

Four Year

The tree has begun to fruit and requires only limited formative pruning. Shorten leaders by one third and prune laterals not required to extend the framework to four buds.

Five Year and Onwards

The tree is established, and should be pruned annually.

Pruning the Cropping Tree

Before pruning, distinguish between spur-bearing varieties, tip-bearing varieties, and an intermediate between the two that bears both on spurs and at the tips. Spur-bearing trees occur more frequently than tip-bearing trees, and they bear most of their fruit yearly at the end of short lateral pieces of wood (spurs) up to about 4 inches long.

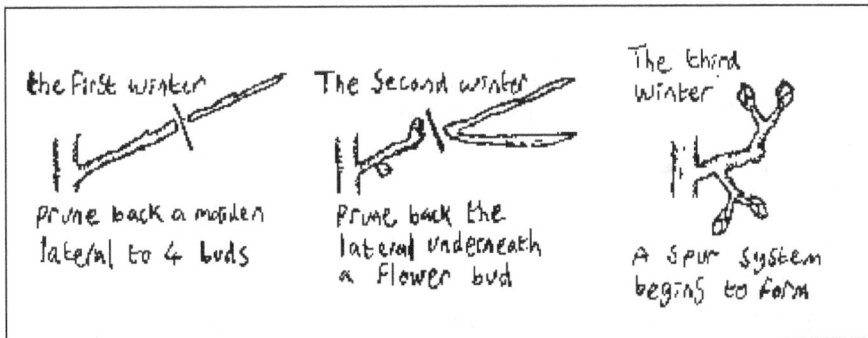

Spur pruning.

Spur-bearing types include apples of the varieties 'Cox's Orange Pippin', 'James Grieve' and 'Sunset', and pears such as 'Conference', 'Doyenne du Commice', and 'Williams Bon Chretien'. Tip-bearers on the other hand produce most of their fruit buds at the tips of slender shoots grown the previous summer, and include the apples 'Worcester Pearmain' and 'Irish Peach', and the pears such as 'Jargonelle' and 'Josephine de Malines'. There are basically three types of pruning that are applied once the main shape of the tree has been established. These are:

- Spur pruning: Spur bearing varieties form spurs naturally, but spur growth can also be induced.

- Renewal pruning: This also depends on the tendency of many apple and pear trees to form flower buds on unpruned two-year-old laterals. It is a technique best used for the strong laterals on the outer part of the tree where there is room for such growth. Pruning long-neglected fruit trees is a task that should be undertaken over a lengthy period, with not more than one third of the branches that require removal being taken each year.

- Regulatory pruning: This is carried out on the tree as a whole, and is aimed at keeping the tree and its environment healthy, e.g., by keeping the centre open so that air can circulate; removing dead or diseased wood; preventing branches from becoming overcrowded (branches should be roughly 50 cm apart and spurs not less than 25 cm apart along the branch framework); and preventing any branches from crossing.

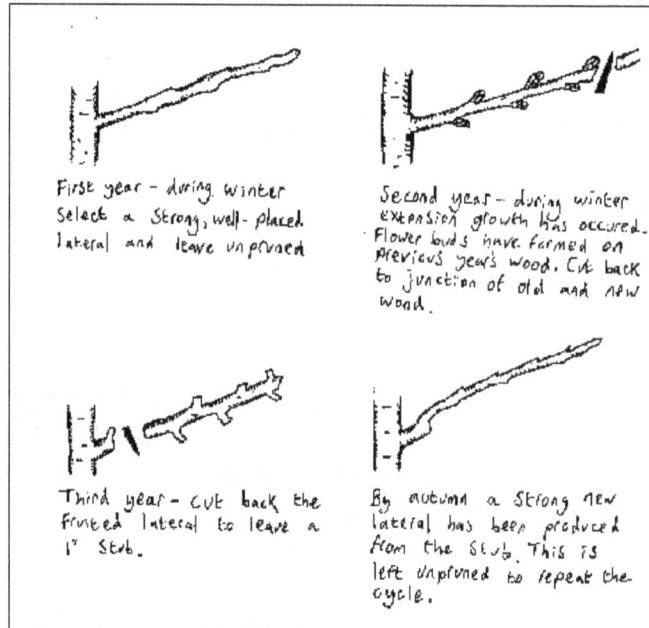

First year – during winter select a strong, well-placed lateral and leave unpruned

Second year – during winter extension growth has occured. Flower buds have formed on previous year's wood. Cut back to junction of old and new wood.

Third year – cut back the fruited lateral to leave a 1" stub.

By autumn a strong new lateral has been produced from the stub. This is left unpruned to repeat the cycle.

Renewal pruning.

Pruning of Tip Bearers

Tip-bearers should be pruned lightly in winter using the regulatory system. Any maiden shoots less than 25 cm in length should be left untouched as they have fruit buds at their tips. Longer shoots are spur pruned to prevent overcrowding and to stimulate the production of more short-tip-bearing shoots the following year. Branch leaders are 'tipped', removing the top three or four buds to a bud facing in the desired direction to make them branch out and so produce more tip-bearing shoots.

Intensive Rotational Grazing

When designing a livestock grazing system to optimize animal health, pasture integrity, and economic productivity, the primary dynamic that must be accounted for is "grazing pressure". Grazing pressure is the closeness to which pasture is grazed, i.e. the amount of grass/forb that is depleted from a pasture in a single grazing event. If the grazing pressure is too high, all three priorities suffer: animal health declines due to the depletion of desirable forage; pasture flora profile and soil structure degrade from overgrazing; and animal productivity decreases due to decreased nutrition. If the grazing pressure is too low, animal productivity (output) will increase because livestock eat as much as they want, but net profit suffers because forage (input) is wasted due to selective grazing. Optimum grazing pressure is achieved when livestock's dietary needs are met and the least amount of forage possible is wasted.

Types of Livestock Grazing Systems

The dynamics of grazing pressure will be affected primarily by the "grazing system" used to manage a pasture. There are two basic types of livestock grazing systems: Continuous and Rotational.

Continuous

Continuous Grazing is the practice of grazing animals continuously on a single pasture for the entire growing season. The primary factor determining grazing pressure in this case is "stocking rate", or the ratio of heads of livestock to acreage. The method entails certain conveniences: no moving animals from pasture to pasture; less monitoring pasture health; and less infrastructure like fences and water.

Rotational

Rotational Grazing, which varies in layout and intensity, is the practice of dividing pasture into two or more cells (called "paddocks"), stocking one at a high rate, and then rotating the livestock to a new pasture once the forage in the present paddock has been grazed to an appropriate degree. The time spent in each paddock is relatively low, but due to the high stocking rates, impact on the pasture is relatively high: soil is trampled; manure is deposited; and forage is grazed rapidly.

Sometimes Rotational Grazing is practiced in such a way that it utilizes less than 7 paddocks and grazing periods of one to two weeks. This method retains some of the benefits of resting periods, but the amount of time spent in the paddocks in relation to the stocking rate may be too long, resulting in overgrazing and pasture destruction. True Rotational Grazing (sometimes called "Intensive Rotational Grazing" to distinguish it) usually utilizes more than 7 (sometimes many) paddocks and grazing periods of between less than one week and half a day. If done carefully, this decreased grazing time/increased stocking rate can ensure that the pasture is not overgrazed, and increase the "impact" of livestock on the pasture.

If designing for optimal animal health, pasture integrity, and economic productivity means designing for optimal grazing pressure dynamics, then there are many situations in which Intensive Rotational Grazing is a superior alternative to Continuous Grazing. Why is this? The answer requires understanding how the naturally occurring pasture ecosystem sustains its own health and the health of grazing herbivores in a natural, unmanaged system.

The Pasture Ecosystem

Pasture ecosystems are evolved to work in coordination with intensive periods of herbivorous grazing. Animals move continuously through open pastures in large herds; they move continuously to find new forage; they herd together into large masses to reduce their individual vulnerability and protect themselves from predators. Thier density in this configuration has a high impact on the pasture. They graze preferred species down to a minimum; their hooves stamp depressions in the ground and upturn flora; and they deposit heavy amounts of urine and manure onto the ground. They must then move on to new pasture because their activity depletes the available forage.

But their intensive migratory grazing habits do more than deplete the pasture. They are also essential to maintaining it as pasture per se. Without these regular intervals of "destruction", which create soil conditions wherein earlier pioneer-type species thrive, plant succession would take it's course and the pasture would steadily progress into a more advanced stage of succession that favors plant species less favorable to the herbivorous diet – woodier, larger volume species with less nutritional concentration.

When animals move through the pasture and interrupt this process of succession, they perform several key functions for maintaining optimal grazing conditions on future visits. Thier grazing of preferred species prevents those species from maturing. Because grasses and forbs posses the most favorable balance between volume and nutritional density at an early stage of maturity, this ensures optimal forage nutrition for later grazing events (albeit not typically within the same herd).

Thier stamping of depressions into the ground and upturning of flora creates the ideal context for the establishment of the seedlings of preferred forage species. One example of this is the fact that some seeds will only germinate when a slight crushing pressure has been applied, for example by a hoof. Another example is that hoof depressions act as tiny microclimates, perfect for germinating seeds. They collect water; they hold the seed below the path of wind, protecting it from drying and cooling exposure; and they provide a slightly warmer ambient temperature than the soil surface.

Lastly, animals promote soil fertility by depositing urine and manure onto the pasture. These depositions comes in very high quantities, which would be detrimental if sustained for a protracted period of time, but because of the punctuated nature of the grazing events, the depositions have time to decompose before another pass is made.

This system is complex and resilient to variation. But it is adapted to work on the basic model of: short, intensive, punctuated grazing events; periodic pasture impaction; forage depletion and fertility deposition; perpetual cycles of early stage succession.

The Problem with Continuous Grazing

The practice of grazing animals continuously on a single pasture for the entire growing season subverts this naturally healthy and sustainable system at multiple levels and as a consequence creates multiple problems for itself, resulting in more inputs and less productivity. Continuous Grazing is inherently susceptible to the following shortcomings, even though variations in continuous grazing management systems exist that are more or less appropriate.

Soil Erosion

While the stamping of depressions and upturning of flora can be a rejuvenating force as a punctuated event, if done continually it will destroy pasture altogether. "Soil" is not dirt. Soil is a living organism, comprised of vast networks other other organisms living in a delicate balance. The existence of soil is premised on the continuity of conditions that preserve these life forms. If livestock are allowed to consume, trample, and foul all of the grasses growing in a pasture, rapidly or slowly, these conditions are threatened. The cycle of grasses growing, dying, decomposing, and regrowing is retarded, and too little of the right kinds of organic matter may be returned to the soil, which will further retard the growth of new grass. If soil is left exposed without a protective layer of grass, elements like wind and rain will directly impact the soil, killing delicate microorganisms which are not adapted to above ground conditions. As microorganisms die and the cycle of plant life is retarded, organic matter steadily disappeares from the soil. Gradually, the soil is reduced to dirt, the dead mineral component of soil, and the ecosystem capable of sustaining healthy forage is lost. Without plant root systems to hold the soil in place, soil structure may degrade so badly that the dirt succombs to eorison and gets carried down the watershed in rain events.

Over Grazing and Spot Grazing

Overgrazing and spot grazing both occur when livestock doesn't have access to the right kind of forage for the right amount of time.

Overgrazing is the depletion of forage to a critically low level due to unsustainable grazing intensity. This can result from stocking rates being too high or the grazing event being too long. Because forage yields in a pasture vary throughout the growing season, grazing intensity will fluctuate (even if the intake of the animals remains the same) and will sometimes be too low, sometimes to high, making it very difficult ensure optimal grazing pressure.

Spot grazing results from the natural and desirable tendency of livestock to seek out the forage they prefer and pass over the ones they don't, called "selective grazing". If this process goes on continually uninterrupted, desirable species will be depleted too badly to recover and undesirable species will flourish, on account both of not being eaten and to having continually depleted competition. In this way the desirable pasture species are gradually weeded out and replaced with undesirable species. This situation can lead to overgrazing, because only the areas that retain desirable forage are grazed while the undesirable sections go ungrazed. Due to the decreasing area of the sections containing desirable forage, the stocking load becomes increasingly too great for the desirable sections and the forage is depleted.

Overgrazing contributes to erosion which can have long term negative impacts on the viability of pasture land for forage. It also directly impacts the health and productivity of livestock due to supplying inadequate nutrition.

Pasture Fouling

Pasture fouling is a similar problem to overgrazing, and also occurs when stocking rates are too high or the grazing event is too long. Except, instead of depleting the pasture of foragable plant species, livestock deposit too much of the ingested forage, or urine and manure, back onto the pasture. If the pasture has no time to rest between deposition events of a certain scale, the soil conditions will become unfavorable to the growth of forage species, due for example to too much nitrogen. As the soil conditions become less favorable, the pasture's ability to break down manure decreases, causing the conditions to become worse, and so on.

In addition to the fouling and loss of available forage, this situation can result in the spread of disease and parasites in livestock. Traditional wisdom says that about one cycle of the moon (~30 days) is the minimum time period of time to ensure that parasites deposited onto the pasture via manure are dead. This is also about how long it will take for manure to decompose to a soil-like state. Continually grazing animals in the same pastures, especially if stocking rates are too high, means continually grazing them in among 30 days worth of manure deposition, which, it seems reasonable to me, makes them roughly 30 times more vulnerable to parasites than they would be if they grazed on a fresh pasture every day.

References

- Natural-building-use-technique: milkwood.net, Retrieved 18 June, 2019
- Wojtkowski, Paul Anthony (2002). Agroecological Perspectives in Agronomy, Forestry, and Agroforestry. Science Publishers. ISBN 978-1-57808-217-9
- Holzer, Sepp (2012). Hügelkultur. Chelsea Green Publishing. pp. 131–134, 139. ISBN 978-1603584647
- Permaculture-strategies-intensive-rotational-grazing: transterraform.com, Retrieved 17 July, 2019

Components of Permaculture

4

- **Folkewall**
- **Composting Toilet**
- **Leaf Mold**
- **Mulch**
- **Three Sisters**
- **Treebog**
- **Spent Mushroom Compost**

Permaculture consists of numerous components such as folkewall, composting toilet, leaf mold, mulch, Three Sisters, treebog, spent mushroom compost, groundcover, etc. This chapter closely examines these key components of permaculture to provide an extensive understanding of the subject.

Folkewall

Folkewall is a hydroponics growing system that is specifically designed to make proper use of limited land to fulfill two important purposes: i.e., purification of greywater and vertical growing technique or living wall. It's basically a vertical growing method and the best example of it is the hanging gardens of the Babylon. Vertical farming helps in efficient use of space and purification of percolating water which may be greywater.

The simplest design of folkewall is a wall of hollow slabs that have proper openings on one or both sides of the wall. You can fill the hollows with inert material like LECA-pebbles (light expanded clay aggregate), gravel, vermiculite or perlite. It's designed to allow the water to travel over the longest path through the wall along with the pebbles. The water is allowed to reach the top in a zig-zag manner in the wall. At this time plant roots grow with the inert material and absorb nutrients directly from the water. Beneficial bacteria grow on the pebbles and break down organic pollutants in the greywater. After consuming the pollutant they release nutrients in the water which are thereafter is absorbed by the plant. A container fixed at the bottom of the wall collect purified water, which can be recycled.

Since herbaceous crops are fast growing crops they are the ideal ones for Folkewall. Perennials that take time to fully grow such as trees or shrubs are not recommended for the Folkewall. Special care should be taken while using the water; avoid using water that contains heavy metals or any unsafe pollutants.

Advantages of Folkewall

- It purifies greywater and is especially helpful in arid climatic conditions.

- It utilizes space efficiently and can be practiced in a green house or in open in frost free climate.

- Water can be reused or recycled an also can be used for other household or irrigation purposes.

- Heat exchanger and temperature buffer in a green house where the wall is combined with greywater purification.

- In hot climate, the living walls give a cooling effect to the building.

Composting Toilet

A composting toilet is a type of dry toilet that treats human excreta by a biological process called composting. This process leads to the decomposition of organic matter and turns human excreta into compost-like material but does not destroy all pathogens. Composting is carried out by microorganisms (mainly bacteria and fungi) under controlled aerobic conditions. Most composting toilets use no water for flushing and are therefore called "dry toilets".

In many composting toilet designs, carbon additives such as sawdust, coconut coir, or peat moss is added after each use. This practice creates air pockets in the human excreta to promote aerobic decomposition. This also improves the carbon-to-nitrogen ratio and reduces potential odor. Most composting toilet systems rely on mesophilic composting. Longer retention time in the composting chamber also facilitates pathogen die-off. The end product can also be moved to a secondary system – usually another composting step – to allow more time for mesophilic composting to further reduce pathogens.

Composting toilets, together with the secondary composting step, produce a humus-like endproduct that can be used to enrich soil if local regulations allow this. Some composting toilets have urine diversion systems in the toilet bowl to collect the urine separately and control excess moisture. A "vermifilter toilet" is a composting toilet with flushing water where earthworms are used to promote decomposition to compost.

Composting toilets do not require a connection to septic tanks or sewer systems unlike flush toilets. Common applications include national parks, remote holiday cottages, ecotourism resorts, off-grid homes and rural areas in developing countries.

Applications

This is the pedestal for a split-system composting toilet
where collection/treatment chambers are located below the bathroom floor.

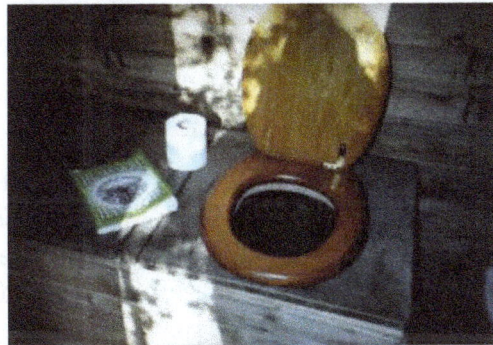

Inexpensive do-it-yourself compost toilet at Dial House, Essex, England,
utilizing an old desk as the toilet unit.

Public composting toilet at a highway rest facility in Sweden.

Composting toilets can be suitable in areas such as a rural area or a park that lacks a suitable water supply, sewers and sewage treatment. They can also help increase the resilience of existing sanitation systems in the face of possible natural disasters such as climate change, earthquakes or tsunami. Composting toilets can reduce or perhaps eliminate the need for a septic tank system to reduce environmental footprint (particularly when used in conjunction with an on-site greywater treatment system).

These types of toilets can be used for resource recovery by reusing sanitized feces and urine as fertilizer and soil conditioner for gardening or ornamental activities.

Basics

Components and Use

A composting toilet consists of two elements: a place to sit or squat and a collection/composting unit. The composting unit consists of four main parts:

- Storage or composting chamber,

- A ventilation unit to ensure that the degradation process in the toilet is predominantly aerobic and to vent odorous gases,

- A leachate collection or urine diversion system to remove excess liquid,

- An access door for extracting the compost.

Many composting toilets collect urine in the same chamber as feces, thus they do not divert urine. Adding small amounts of water that is used for anal cleansing is no problem for the composting toilet to handle.

Some composting toilets divert urine (and water used for anal washing) to prevent the creation of anaerobic conditions that can result from over saturation of the compost, which leads to odors and vector problems. This usually requires all users to use the toilet in a seated position. Offering a waterless urinal in addition to the toilet can help keep excess amounts of urine out of the composting chamber. Alternatively, in rural areas, men and boys may be encouraged just to find a tree.

Construction

The composting chamber can be constructed above or below ground level. It can be inside a structure or include a separate superstructure.

A drainage system removes leachate. Otherwise, excess moisture can cause anaerobic conditions and impede decomposition. Urine diversion can improve compost quality, since urine contains large amounts of ammonia that inhibits microbiological activity.

Composting toilets greatly reduce human waste volumes through psychrophilic, thermophilic or mesophilic composting. Keeping the composting chamber insulated and warm protects the composting process from slowing due to low temperatures.

Odorous Gases

The following gases may be emitted during the composting process that takes place in composting toilets: hydrogen sulfide (H_2S), ammonia, nitrous oxide (N_2O) and volatile organic compounds (VOCs). These gases can potentially lead to complaints about odours. Some methane may also be present, but it is not odorous.

Pathogen Removal

Excreta-derived compost recycles fecal nutrients, but it can carry and spread pathogens if the process of reuse of excreta is not done properly. Pathogen destruction rates in composting toilets are usually low, particularly helminth eggs, such as Ascaris eggs. This carries the risk of spreading disease if a proper system management is not in place. Compost from human excreta processed under only mesophilic conditions or taken directly from the compost chamber is not safe for food production. High temperatures or long composting times are required to kill helminth eggs, the hardiest of all pathogens. Helminth infections are common in many developing countries.

In thermophilic composting bacteria that thrive at temperatures of 40–60 °C (104–140 °F) oxidize (break down) waste into its components, some of which are consumed in the process, reducing volume and eliminating potential pathogens. To destroy pathogens, thermophilic composting must heat the compost pile sufficiently, or enough time (1–2 years) must elapse since fresh material was added that biological activity has had the same pathogen removal effect.

One guideline claims that pathogen levels are reduced to a safe level by thermophilic composting at temperatures of 55 °C for at least two weeks or at 60 °C for one week. An alternative guideline claims that complete pathogen destruction may be achieved already if the entire compost heap reaches a temperature of 62 °C (144 °F) for one hour, 50 °C (122 °F) for one day, 46 °C (115 °F) for one week or 43 °C (109 °F) for one month, although others regard this as overly optimistic.

Design Considerations

Composting toilet with a seal in the lid in Germany.

Environmental Factors

Four main factors affect the decomposition process:

- Sufficient oxygen is necessary for aerobic composting.

- Moisture content from 45 to 70 percent (heuristically, "the compost should feel damp to the touch, with only a drop or two of water expelled when tightly squeezed in the hand").

- Temperature between 40 and 50 °C (achieved through proper chamber dimensioning and possibly active mixing).

- Carbon-to-nitrogen ratio (C:N) of 25:1.

Additives and Bulking Material

Human excreta and food waste do not provide optimum conditions for composting. Usually the water and nitrogen content is too high, particularly when urine is mixed with feces. Additives or "bulking material", such as wood chips, bark chips, sawdust, shredded dry leaves, ash and pieces of paper can absorb moisture. The additives improve pile aeration and increase the carbon to nitrogen ratio. Bulking material also covers feces and reduces insect access. Absent sufficient bulking material, the material may become too compact and form impermeable layers, which leads to anaerobic conditions and odour.

Leachate Management

Leachate removal controls moisture levels, which is necessary to ensure rapid, aerobic composting. Some commercial units include a urine-separator or urine-diverting system and a drain at the bottom of the composter for this purpose.

Aeration and Mixing

Microbial action also requires oxygen, typically from the air. Commercial systems provide ventilation that moves air from the bathroom, through the waste container, and out a vertical pipe, venting above the roof. This air movement (via convection or fan forced) passes carbon dioxide and odors.

Some units require manual methods for periodic aeration of the solid mass such as rotating the composting chamber or pulling an "aerator rake" through the mass.

Types

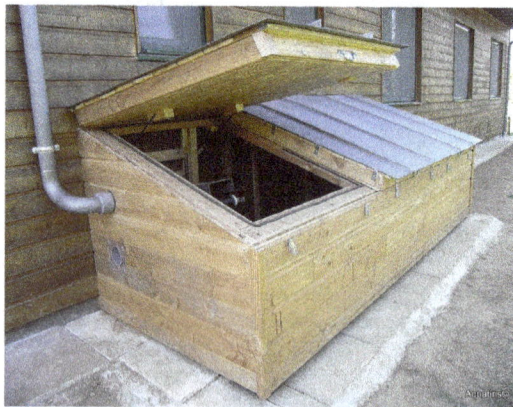

External composting chamber of a composting toilet at a house in France.

Commercial units and construct-it-yourself systems are available. Variations include number of composting vaults, removable vault, urine diversion and active mixing/aeration.

Slow Composting Toilets

Most composting toilets use slow composting which is also called "cold composting". The compost heap is built up step by step over time.

The finished end product from "slow" composting toilets ("moldering toilets" or "moldering priv-ies" in the US), is generally not free of pathogens. World Health Organization Guidelines from 2006 offer a framework for safe reuse of excreta, using a multiple barrier approach.

Slow composting toilets employ a passive approach. Common applications involve modest and often seasonal use, such as remote trail networks. They are typically designed such that the ma-terials deposited can be isolated from the operational part. The toilet can also be closed to allow further mesophilic composting. Slow composting toilets rely on long retention times for pathogen reduction and for decomposition of excreta or on the combination of time and the addition of red wriggler worms for vermi-composting. Worms can be introduced to accelerate composting. Some jurisdictions of the US consider these worms as invasive species.

Active Composters

"Self-contained" composting toilets compost in a container within the toilet unit. They are slightly larger than a flush toilet, but use roughly the same floor space. Some units use fans for aeration, and optionally, heating elements to maintain optimum temperatures to hasten the composting process and to evaporate urine and other moisture. Operators of composting toilets commonly add a small amount of absorbent carbon material (such as untreated sawdust, coconut coir, peat moss) after each use to create air pockets to encourage aerobic processing, to absorb liquid and to create an odor barrier. This additive is sometimes referred to as "bulking agent". Some owner-operators use microbial "starter" cultures to ensure composting bacteria are in the process, although this is not critical.

Vermifilter Toilet

A "vermifilter toilet" is a composting toilet with flushing water where earthworms are used to pro-mote decomposition to compost. It can be connected to a low-flush or a micro-flush toilet which uses about 500 millilitres (17 US fl oz) per use. Solids accumulate on the surface of the filter bed while liquid drains through the filter medium and is discharged from the reactor. The solids (feces and toilet paper) are aerobically digested by aerobic bacteria and composting earthworms into castings (humus), thereby significantly reducing the volume of organic material.

Other

Some units employ roll-away containers fitted with aerators, while others use sloped-bottom tanks.

Maintenance

Maintenance is critical to ensure proper operation, including odor prevention. Maintenance tasks include: cleaning, servicing technical components such as fans and removal of compost, leachate and urine. Urine removal is only required for those types of composting toilets using urine diversion.

Once composting is complete (or more often), the compost must be removed from the unit. How often this occurs is a function of container size, usage and composting conditions, such as tempera-ture. Active, hot composting may span months only while passive, cold composting may require years. Properly managed units yield output volumes of about 10% of inputs.

Uses of Compost

Finished compost from a composting toilet ready for application as
soil improvement in Kiel-Hassee, Germany.

The material from composting toilets is a humus-like material, which can be suitable as a soil amendment for agriculture. Compost from residential composting toilets can be used in domestic gardens, and this is the main such use.

Enriching soil with compost adds substantial nitrogen, phosphorus, potassium, carbon and calcium. In this regard compost is equivalent to many fertilizers and manures purchased in garden stores. Compost from composting toilets has a higher nutrient availability than the dried feces that result from a urine-diverting dry toilet.

Urine is typically present, although some is lost via leaching and evaporation. Urine can contain up to 90 percent of the residual nitrogen, up to 50 percent of the phosphorus, and up to 70 percent of the potassium.

Compost derived from these toilets has in principle the same uses as compost derived from other organic waste products, such as sewage sludge or municipal organic waste. However, users of excreta-derived compost must consider the risk of pathogens.

Pharmaceutical Residues

Excreta-derived compost may contain prescription pharmaceuticals. Such residues are also present in conventional wastewater treatment effluent. This could contaminate groundwater. Among the medications that have been found in groundwater in recent years are antibiotics, antidepressants, blood thinners, ACE inhibitors, calcium-channel blockers, digoxin, estrogen, progesterone, testosterone, Ibuprofen, caffeine, carbamazepine, fibrates and cholesterol-reducing medications. Between 30% and 95% of pharmaceuticals medications are excreted by the human body. Medications that are lipophilic (dissolved in fats) are more likely to reach groundwater by leaching from fecal wastes. Wastewater treatment plants remove an average of 60% of these medications. The percentage of medications degraded during composting of excreta has not yet been reported.

Comparison

Pit Latrines

Unlike pit latrines, composting toilets convert feces into a dry, odorless material, avoiding the issues surrounding liquid fecal sludge management (e.g. odor, insects and disposal). These toilets minimize the risk of water pollution through the safe containment of feces in above-ground vaults, which allows the toilets to be sited in locations where pit-based systems are not appropriate.

However, composting toilets face higher capital costs (although lifecycle costs might be lower) and greater complexity (for instance, adding covering materials and managing moisture content).

Flush Toilets

Unlike flush toilets, composting toilets do not dilute excreta and create wastewater streams which must be treated before disposal. On the other hand, wastewater treatment plants can centralize waste management for an entire community, with potentially greater efficiency.

Urine-Diverting Dry Toilets

Composting toilets are more difficult to maintain than other types of dry toilets, like urine-diverting dry toilets (UDDT) with which they are often confused. This is due to the need to maintain a consistent and relatively high moisture content, as well as the relatively high complexity of composting toilets compared to UDDTs. Apart from that, composting toilets are quite similar to UDDTs, sharing many of the same advantages and disadvantages.

Finland

Numerous sparsely settled villages in rural areas in Finland are not connected to municipal water supply or sewer networks, requiring homeowners to operate their own systems. Individual private wells, i.e. shallow dug wells or boreholes in the bedrock, are often used for water supply, and many homeowners have opted for composting toilets. In addition, these toilets are common at holiday homes, often located near sensitive water bodies. For these reasons, many manufacturers of composting toilets are based in Finland, including Biolan, Ekolet, Kekkilä, Pikkuvihreä and Raita Environment.

Estimates made by leading Finnish composting toilet manufacturers and the Global Dry Toilet Association of Finland provided the following 2014 figures for composting toilet use in Finland:

- About 4% of single-family homes not connected to a public sewer network are equipped with a composting toilet.

- Some 200,000 manufactured composting toilets are thought to serve holiday homes, matched by the number of other dry toilets. The simplest ones are sited in an outhouse.

Germany

Composting container of "TerraNova" composting toilet, showing open
removal chamber (town house at the ecological settlement
Hamburg-Allermöhe, Germany).

Composting toilets have been successfully installed in houses with up to four floors. An estimate from 2008 put the number of composting toilets in households in Germany at 500. Most of these residences are also connected to a sewer system; the composting toilet was not installed due to a lack of sewer system but for other reasons, mainly because of an "ecological mindset" of the owners.

In Germany and Austria, composting toilets and other types of dry toilets have been installed in single and multi-family houses (e.g. Hamburg, Freiburg, Berlin), ecological settlements (e.g. Hamburg-Allermöhe, Hamburg-Braamwisch, Kiel-Hassee, Bielefeld-Waldquelle, Wien-Gänserndorf) and in public buildings (e.g. Ökohaus Rostock, VHS-Ökostation Stuttgart-Wartberg, public toilets in recreational areas, restaurants and huts in the Alps, house boats and forest Kindergartens).

The ecological settlement in Hamburg-Allermöhe has had composting toilets since 1982. The settlement of 36 single-family houses with approximately 140 inhabitants uses composting toilets, rainwater harvesting and constructed wetlands. Composting toilets save about 40 litres of water per capita per day compared to a conventional flush toilet (10 liter per flush), which adds up to 2,044 m³ water savings per year for the whole settlement.

United States

Slow composting toilets have been installed by the Green Mountain Club in Vermont's woodlands. They employ multiple vaults (called cribs) and a movable building. When one of the vaults fills, the building is moved over an empty vault. The full vault is left untouched for as long as possible (up to three years) before it is emptied. The large surface area and exposure to air currents can cause the pile to dry out. To counteract this, signs instruct users to urinate in the toilet. The club also uses pit latrines and simple bucket toilets with woodchips and external composting and directs users to urinate in the forest to prevent odiferous anaerobic conditions.

Worldwide

Composting toilets with a large composting container (of the type Clivus Multrum and derivations of it) are popular in US, Canada, Australia, New Zealand and Sweden. They can be bought and installed as commercial products, as designs for self builders or as "design derivatives" which are marketed under various names. It has been estimated that approximately 10,000 such toilets might be in use worldwide.

Leaf Mold

Leaf mold (Leaf mould outside of the United States) is the product of slow decomposition of deciduous shrub and tree leaves. It is a form of compost produced primarily by fungal breakdown.

Leaves shed in autumn tend to have a very low nitrogen content and are often dry. Their main constituents are cellulose and lignin. Because of this, autumn leaves break down far more slowly than most other compost ingredients with very little bacterial decomposition involved.

Time and Process

Fungal decomposition of a heap of leaves in the open can take between one and two years to break down into a dark brown fine powdery humic matter. During the two to three years that the process takes to complete in damp temperate climates, a succession of different fungal species may be involved. A range of micro detritivores are also involved in converting the leaf material into a fine-grained humus, including many isopods, millipedes, earthworms, etc.

Uses

In the natural environment the slow decomposition of leaves provides a moist growing medium for young plants and also protects the ground from drying out during periods of low rainfall. It is a significant component of soil organic matter, particularly in temperate deciduous woodland. The slow rate of decomposition allows the plant nutrients bound up in the leaves to be released slowly back into the environment where they can be re-used by plants. Autumn leaves are often collected as part of gardening or farming and kept in pits or containers so that the leaf mold can be used in the garden. The presence of oxygen from the air and sufficient moisture are essential for leaf decomposition. Leaf mold is not high in nutrient content, but it is excellent humic soil conditioner because of its ability to retain moisture and provide a good growing medium for seedling roots. Leaves collected off roads and pavements may be contaminated by pollutants which can become more concentrated as the leaves decompose into a smaller volume.

Mulch

A mulch is a layer of material applied to the surface of soil. Reasons for applying mulch include conservation of soil moisture, improving fertility and health of the soil, reducing weed growth and enhancing the visual appeal of the area.

A mulch is usually, but not exclusively, organic in nature. It may be permanent (e.g. plastic sheeting) or temporary (e.g. bark chips). It may be applied to bare soil or around existing plants. Mulches of manure or compost will be incorporated naturally into the soil by the activity of worms and other organisms. The process is used both in commercial crop production and in gardening, and when applied correctly, can dramatically improve soil productivity.

Uses

Many materials are used as mulches, which are used to retain soil moisture, regulate soil temperature, suppress weed growth, and for aesthetics. They are applied to the soil surface, around trees, paths, flower beds, to prevent soil erosion on slopes, and in production areas for flower and vegetable crops. Mulch layers are normally 2 inches (5.1 cm) or more deep when applied.

They are applied at various times of the year depending on the purpose. Towards the beginning of the growing season, mulches serve initially to warm the soil by helping it retain heat which is lost during the night. This allows early seeding and transplanting of certain crops, and encourages faster growth. As the season progresses, mulch stabilizes the soil temperature and moisture, and prevents the growing of weeds from seeds. In temperate climates, the effect of mulch is dependent upon the time of year they are applied and when applied in fall and winter, are used to delay the growth of perennial plants in the spring or prevent growth in winter during warm spells, which limits freeze thaw damage.

The effect of mulch upon soil moisture content is complex. Mulch forms a layer between the soil and the atmosphere preventing sunlight from reaching the soil surface, thus reducing evaporation. However, mulch can also prevent water from reaching the soil by absorbing or blocking water from light rains.

In order to maximise the benefits of mulch, while minimizing its negative influences, it is often applied in late spring/early summer when soil temperatures have risen sufficiently, but soil moisture content is still relatively high.

Plastic mulch used in large-scale commercial production is laid down with a tractor-drawn or stand-alone layer of plastic mulch. This is usually part of a sophisticated mechanical process, where raised beds are formed, plastic is rolled out on top, and seedlings are transplanted through it. Drip irrigation is often required, with drip tape laid under the plastic, as plastic mulch is impermeable to water.

Materials

Rubber mulch nuggets in a playground. The white fibers are nylon cords, which are present in the tires from which the mulch is made.

Materials used as mulches vary and depend on a number of factors. Use takes into consideration availability, cost, appearance, the effect it has on the soil—including chemical reactions and pH, durability, combustibility, rate of decomposition, how clean it is—some can contain weed seeds or plant pathogens.

Shredded wood used as mulch. This type of mulch is often dyed to improve its appearance in the landscape.

A variety of materials are used as mulch:

- Organic residues: Grass clippings, leaves, hay, straw, kitchen scraps comfrey, shredded bark, whole bark nuggets, sawdust, shells, woodchips, shredded newspaper, cardboard, wool, animal manure, etc. Many of these materials also act as a direct composting system, such as the mulched clippings of a mulching lawn mower, or other organics applied as sheet composting.

- Compost: Fully composted materials are used to avoid possible phytotoxicity problems. Materials that are free of seeds are ideally used, to prevent weeds being introduced by the mulch.

- Old carpet (synthetic or natural): Makes a free, readily available mulch. Rubber mulch: made from recycled tire rubber.

- Plastic mulch: Crops grow through slits or holes in thin plastic sheeting. This method is predominant in large-scale vegetable growing, with millions of acres cultivated under plastic mulch worldwide each year (disposal of plastic mulch is cited as an environmental problem).

- Rock and gravel can also be used as a mulch. In cooler climates the heat retained by rocks may extend the growing season.

Pine needles used as mulch. Also called "pinestraw" in the southern US.

In some areas of the United States, such as central Pennsylvania and northern California, mulch is often referred to as "tanbark", even by manufacturers and distributors. In these areas, the word "mulch" is used specifically to refer to very fine tanbark or peat moss.

Aged Compost mulch on a flower bed.

Organic Mulches

Organic mulches decay over time and are temporary. The way a particular organic mulch decomposes and reacts to wetting by rain and dew affects its usefulness.

Mulching coconut farm.

Some mulches such as straw, peat, sawdust and other wood products may for a while negatively affect plant growth because of their wide carbon to nitrogen ratio, because bacteria and fungi that decompose the materials remove nitrogen from the surrounding soil for growth. However, whether this effect has any practical impact on gardens is disputed by researchers and the experience of gardeners. Organic mulches can mat down, forming a barrier that blocks water and air flow between the soil and the atmosphere. Vertically applied organic mulches can wick water from the soil to the surface, which can dry out the soil. Mulch made with wood can contain or feed termites, so care must be taken about not placing mulch too close to houses or building that can be damaged by those insects. Some mulch manufacturers recommend putting mulch several inches away from buildings.

Commonly available organic mulches include:

Leaves

- Leaves from deciduous trees, which drop their foliage in the autumn/fall. They tend to be dry and blow around in the wind, so are often chopped or shredded before application. As they decompose they adhere to each other but also allow water and moisture to seep down to the soil surface. Thick layers of entire leaves, especially of maples and oaks, can form a soggy mat in winter and spring which can impede the new growth lawn grass and other plants. Dry leaves are used as winter mulches to protect plants from freezing and thawing in areas with cold winters; they are normally removed during spring.

Grass Clippings

- Grass clippings from mowed lawns are sometimes collected and used elsewhere as mulch. Grass clippings are dense and tend to mat down, so are mixed with tree leaves or rough compost to provide aeration and to facilitate their decomposition without smelly putrefaction. Rotting fresh grass clippings can damage plants; their rotting often produces a damaging buildup of trapped heat. Grass clippings are often dried thoroughly before application, which mediates against rapid decomposition and excessive heat generation. Fresh green grass clippings are relatively high in nitrate content, and when used as a mulch, much of the nitrate is returned to the soil, conversely the routine removal of grass clippings from the lawn results in nitrogen deficiency for the lawn.

Peat Moss

- Peat moss or sphagnum peat, is long lasting and packaged, making it convenient and popular as a mulch. When wetted and dried, it can form a dense crust that does not allow water to soak in. When dry it can also burn, producing a smoldering fire. It is sometimes mixed with pine needles to produce a mulch that is friable. It can also lower the pH of the soil surface, making it useful as a mulch under acid loving plants.

However peat bogs are a valuable wildlife habitat, and peat is also one of the largest stores of carbon (in Britain, out of a total estimated 9952 million tonnes of carbon in British vegetation and soils, 6948 million tonnes carbon are estimated to be in Scottish, mostly peatland, soils), so gardeners who wish to protect the environment will choose more sustainable alternatives.

Arborist Wood Chips

- Wood chips are a byproduct of the pruning of trees by arborists, utilities and parks; they are used to dispose of bulky waste. Tree branches and large stems are rather coarse after chipping and tend to be used as a mulch at least three inches thick. The chips are used to conserve soil moisture, moderate soil temperature and suppress weed growth. The decay of freshly produced chips from recently living woody plants, consumes nitrate; this is often off set with a light application of a high-nitrate fertilizer. Wood chips are most often used under trees and shrubs. When used around soft stemmed plants, an unmulched zone is left around the plant stems to prevent stem rot or other possible diseases. They are often used to mulch trails, because they are readily produced with little additional cost outside of the normal disposal cost of tree maintenance. Wood chips come in various colors.

Woodchip Mulch

- Woodchip mulch is a byproduct of reprocessing used (untreated) timber (usually packaging pallets), to dispose of wood waste by creating woodchip mulch. The chips are used to conserve soil moisture, moderate soil temperature and suppress weed growth. Woodchip mulch is often used under trees, shrubs or large planting areas and can last much longer than arborist mulch. In addition, many consider woodchip mulch to be visually appealing, as it comes in various colors. Woodchips can also be reprocessed into playground woodchip to be used as an impact-attenuating playground surfacing.

Bark Chips

Bark chips.

- Bark chips of various grades are produced from the outer corky bark layer of timber trees. Sizes vary from thin shredded strands to large coarse blocks. The finer types are very attractive but have a large exposed surface area that leads to quicker decay. Layers two or three inches deep are usually used, bark is relativity inert and its decay does not demand soil nitrates. Bark chips are also available in various colors.

Straw Mulch

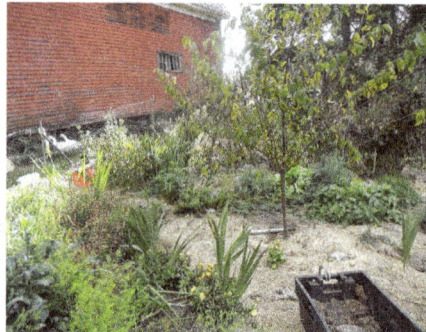

Permaculture garden with a fruit tree, herbs, flowers and vegetables mulched with hay.

- Straw mulch or field hay or salt hay are lightweight and normally sold in compressed bales. They have an unkempt look and are used in vegetable gardens and as a winter covering. They are biodegradable and neutral in pH. They have good moisture retention and weed controlling properties but also are more likely to be contaminated with weed seeds. Salt hay is less likely to have weed seeds than field hay. Straw mulch is also available in various colors.

Pine Straw

- The needles that drop from pine trees is termed *pine straw*. It is available in bales. Pine straw has an attractive look and is used in landscape and garden settings. On application pine needles tend to weave together, a characteristic that helps the mulch hold stormwater on steeper slopes. This interlocking tendency combined with a resistance to floating gives it further advantages in maintaining cover and preventing soil erosion. The interlocking tendency also helps keep the mulch structure from collapsing and forming a barrier to infiltration. Pine straw is reputed to create ideal conditions for acid-loving plants. Pine straw may help to acidify soils but studies indicate this effect is often too small to be measurable. Cardboard/newspaper.

- Cardboard or newspaper can be used as mulches. These are best used as a base layer upon which a heavier mulch such as compost is placed to prevent the lighter cardboard/newspaper layer from blowing away. By incorporating a layer of cardboard/newspaper into a mulch, the quantity of heavier mulch can be reduced, whilst improving the weed suppressant and moisture retaining properties of the mulch. However, additional labour is expended when planting through a mulch containing a cardboard/newspaper layer, as holes must be cut for each plant. Sowing seed through mulches containing a cardboard/newspaper layer is impractical. Application of newspaper mulch in windy weather can be facilitated by briefly pre-soaking the newspaper in water to increase its weight.

Carpet

- Synthetic carpet that is composed of artificial fibers may be removed after planting to prevent fibers taking a long time to decompose, whereas carpet made from natural fibers may be kept in place, blocking competition from weeds. Rain is absorbed by carpet and then slowly released into the soil, reducing watering needs.

Colored Mulch

Some organic mulches are colored red, brown, black, and other colors. Isopropanolamine, specifically 1-Amino-2-propanol or DOW monoisopropanolamine, may be used as a pigment dispersant and color fastener in these mulches. Types of mulch which can be dyed include: wood chips, bark chips (barkdust) and pine straw. Colored mulch is made by dyeing the mulch in a water-based solution of colorant and chemical binder. When colored mulch first entered the market, most formulas were suspected to contain toxic, heavy metals and other contaminates. Today, "current investigations indicate that mulch colorants pose no threat to people, pets or the environment. The dyes currently used by the mulch and soil industry are similar to those used in the cosmetic and other manufacturing industries (i.e., iron oxide)," as stated by the Mulch and Soil Council. Colored mulch can be applied anywhere non-colored mulch is used (such as large bedded areas or around plants) and features many of the same gardening benefits as traditional mulch, such as improving soil productivity and retaining moisture. As mulch decomposes, just as with non-colored mulch, more mulch may need to be added to continue providing benefits to the soil and plants. However, if mulch is faded, spraying dye to previously spread mulch in order to restore color is an option.

Anaerobic Mulch

Mulch normally smells like freshly cut wood, but sometimes develops a toxicity that causes it to smell like vinegar, ammonia, sulfur or silage. This happens when material with ample nitrogen content is not rotated often enough and it forms pockets of increased decomposition. When this occurs, the process may become anaerobic and produce these phytotoxic materials in small quantities. Once exposed to the air, the process quickly reverts to an aerobic process, but these toxic materials may be present for a period of time. If the mulch is placed around plants before the toxicity has had a chance to dissipate, then the plants could very likely be damaged or killed depending on their hardiness. Plants that are predominantly low to the ground or freshly planted are the most susceptible, and the phytotoxicity may prevent germination of some seeds.

If sour mulch is applied and there is plant kill, the best thing to do is to water the mulch heavily. Water dissipates the chemicals faster and refreshes the plants. Removing the offending mulch may have little effect, because by the time plant kill is noticed, most of the toxicity is already dissipated. While testing after plant kill will not likely turn up anything, a simple pH check may reveal high acidity, in the range of 3.8 to 5.6 instead of the normal range of 6.0 to 7.2. Finally, placing a bit of the offending mulch around another plant to check for plant kill will verify if the toxicity has departed. If the new plant is also killed, then sour mulch is probably not the problem.

Groundcovers

Groundcovers are plants which grow close to the ground, under the main crop, to slow the development of weeds and provide other benefits of mulch. They are usually fast-growing plants that continue growing with the main crops. By contrast, cover crops are incorporated into the soil or killed with herbicides. However, live mulches also may need to be mechanically or chemically killed eventually to prevent competition with the main crop.

Some groundcovers can perform additional roles in the garden such as nitrogen fixation in the case of clovers, dynamic accumulation of nutrients from the subsoil in the case of creeping comfrey (*Symphytum ibericum*), and even food production in the case of Rubus tricolor.

On-Site Production

Owing to the great bulk of mulch which is often required on a site, it is often impractical and expensive to source and import sufficient mulch materials. An alternative to importing mulch materials is to grow them on site in a "mulch garden" – an area of the site dedicated entirely to the production of mulch which is then transferred to the growing area. Mulch gardens should be sited as close as possible to the growing area so as to facilitate transfer of mulch materials.

Mulching over Unwanted Plants

Sufficient mulch over plants will destroy them, and may be more advantageous than using herbicide, cutting, mowing, pulling, raking, or tilling. The higher the temperature that this "mulch" is composted, the quicker the reduction of undesirable materials. "Undesirable materials" may include living seed, plant "trash", as well as pathogens such as from animal feces, urine (e.g. hantavirus), fleas, lice, ticks, etc.

In some ways this improves the soil by attracting and feeding earthworms, and adding humus. Earthworms "till" the soil, and their feces are among the best fertilizers and soil conditioners.

Urine may be toxic to plants if applied to growing areas undiluted.

Polypropylene and Polyethylene Mulch

Polypropylene mulch is made up of polypropylene polymers where polyethylene mulch is made up of polyethylene polymers. These mulches are commonly used in many plastics. Polyethylene is used mainly for weed reduction, where polypropylene is used mainly on perennials. This mulch is placed on top of the soil and can be done by machine or hand with pegs to keep the mulch tight against the soil. This mulch can prevent soil erosion, reduce weeding, conserve soil moisture, and increase temperature of the soil. Ultimately this can reduce the amount of work a farmer may have to do, and the amount of herbicides applied during the growing period. The black and clear mulches capture sunlight and warm the soil increasing the growth rate. White and other reflective colours will also warm the soil, but they do not suppress weeds as well. This mulch may require other sources of obtaining water such as drip irrigation since it can reduce the amount of water that reaches the soil. This mulch needs to be manually removed at the end of the season since when it starts to break down it breaks down into smaller pieces. If the mulch is not removed before it starts to break down eventually it will break down into ketones and aldehydes polluting the soil. This mulch is technically biodegradable but does not break down into the same materials the more natural biodegradable mulch does.

Biodegradable Mulch

Quality biodegradable mulches are made out of plant starches and sugars or polyester fibers. These starches can come from plants such as wheat and corn. These mulch films may be a bit more permeable allowing more water into the soil. This mulch can prevent soil erosion, reduce weeding, conserve soil moisture, and increase temperature of the soil. Ultimately this can reduce the amount of herbicides used and manual labor farmers may have to do throughout the growing season. At the end of the season these mulches will start to break down from heat. Microorganisms in the soil break down the mulch into two components, water and CO_2, leaving no toxic residues behind. This source of mulch is even less manual labor since it does not need to be removed at the end of the season and can actually be tilled into the soil. With this mulch it is important to take into consideration that it's much more delicate than other kinds. It should be placed on a day which is not too hot and with less tension than other synthetic mulches. These also can be placed by machine or hand but it is ideal to have a more starchy mulch that will allow it to stick to the soil better.

Three Sisters

The Three Sisters are the three main agricultural crops of various indigenous groups in the Americas: winter squash, maize (corn), and climbing beans (typically tepary beans or common beans). Originating in Mesoamerica, these three crops were carried northward, up the river valleys over generations of time, far afield to the Mandan and Iroquois who, among others, used these "Three Sisters" as trade goods.

In a technique known as companion planting the three crops are planted close together. Flat-topped mounds of soil are built for each cluster of crops. Each mound is about 30 cm (12 in) high and 50 cm (20 in) wide, and several maize seeds are planted close together in the center of each mound. In parts of the Atlantic Northeast, rotten fish or eels are buried in the mound with the maize seeds, to act as additional fertilizer where the soil is poor. When the maize is 15 cm (6 inches) tall, beans and squash are planted around the maize, alternating between the two kinds of seeds. The process to develop this agricultural knowledge took place over 5,000–6,500 years. Squash was domesticated first, with maize second and then beans being domesticated. Squash was first domesticated 8,000–10,000 years ago.

The three crops benefit from each other. The maize provides a structure for the beans to climb, eliminating the need for poles. The beans provide the nitrogen to the soil that the other plants use, and the squash spreads along the ground, blocking the sunlight, helping prevent the establishment of weeds. The squash leaves also act as a "living mulch", creating a microclimate to retain moisture in the soil, and the prickly hairs of the vine deter pests. Corn, beans, and squash contain complex carbohydrates, essential fatty acids and all nine essential amino acids.

Native Americans throughout North America are known for growing variations of Three Sisters gardens. The milpas of Mesoamerica are farms or gardens that employ companion planting on a larger scale. The Ancestral Puebloans are known for adopting this garden design in a drier environment. The Tewa and other peoples of the North American Southwest often included a "fourth Sister", Rocky Mountain bee plant (*Cleome serrulata*), which attracts bees to help pollinate the beans and squash.

Cahokian, Mississippian and Mvskoke Culture

Corn, squash and beans were planted ca. 800 AD in the largest Native American city north of the Rio Grande known as Cahokia, in what is now known as the US state of Illinois, across the river from St Louis, Missouri. The Three Sisters crops were responsible for the surplus food that created an expanded population throughout the extended Mississippi River valley and tributaries, creating the Mississippian and Mvskoke cultures that flourished from ca. 800 ce to ca. 1600 when physical contact with Spanish explorers brought European disease, death, and cultural collapse.

Iroquois Culture

Among the Haudenosaunee, notably the Seneca, women were responsible for crop cultivation, including the 'Three Sisters'. Men had more cause to travel for extended periods of time, such as for hunting expeditions, diplomatic missions or the tribe's numerous wars with nearby rivals. However, men took part in the initial preparation for the planting of the 'Three Sisters' by clearing the planting ground. After a sufficient area of soil was prepared, groups of women (related to each other) took on all the planting, weeding, and harvesting.

Maya Culture

Maya diet focused on three domesticated crops (staple crops): maize, squash, and beans (typically *Phaseolus vulgaris*). Among the three, maize was the central component of the diet of the ancient Maya, and figured prominently in Maya mythology and ideology. Archaeological evidence suggest that Chapalote-Nal-Tel was the dominant species, however it is likely others were being exploited also. Maize was used and eaten in a variety of ways, but was always nixtamalized.

Treebog

A treebog is a type of low-tech compost toilet. It consists of a raised platform above a compost pile surrounded by densely planted willow trees or other nutrient-hungry vegetation. It can be considered an example of permaculture design, as it functions as a system for converting urine and feces to biomass, without the need to handle excreta.

The treebog is a simple method of composting wastes. Abrahams claims that from 1995-2011, around 1500 treebogs may have been built in Britain. They have been on sites ranging from fruit farms, pick-your-own enterprises, campsites, an angling lake, festival sites, remote/low impact dwellings, holiday cottages, allotments, and church yards where there is no mains water supply.

In 2011, Abrahams claimed that the treebog had attracted the attention of NGOs and aid workers who hope to develop its potential for shanty towns or refugee camps - anywhere that water is scarce and the population pressure on resources is high.

Plant Growth

A treebog is simply a controlled compost heap whose function has been enhanced by use of moisture or nutrient-hungry trees. They use no water, purify waste as they create a biomass resource, and also contain the organic waste material, thus preventing the spread of disease.

The main requirement is that the planted species should be nutrient-hungry. It is a bonus if they can be harvested or coppiced for productive uses, e.g. willow cultivars. Apart from willow coppice, soft fruit such as black currants and sweet-smelling herbs such as mint will thrive around a treebog. If left unmanaged, a treebog will soon be surrounded by weed species such as nettles, but a little management and conscious planting can create a fertile and productive bog garden.

Both the solids and liquids are deposited within the treebog base, where the solids compost and the liquids soak through the soil. The associated dense root zone enables the nitrogen to be rapidly absorbed and metabolized by the mycorhyzal species. The feces are contained within the treebog base, which is well ventilated to allow aerobic decomposition to occur, the mineralized material feeding the trees around it.

Construction

A seating platform/cubicle is mounted at least one meter high. The area beneath the seating platform is enclosed by a double-layer of chicken wire; this acts as an effective child-proof barrier and allows air to circulate through the compost heap.This allows for optimum plant growing conditions.

Sawdust, straw, woodchip, ash or other high-carbon matter is used to balance the high-nitrogen content of the urine. One design used Effective Micro-organism bran, which helped keep the treebog virtually odour free.

The space between the wire is stuffed with straw, which acts as a wick to help sop up excess urine, preventing the likelihood of odour problems due to incomplete biological absorption of the nitrogen from the urine. The straw-filled wire also enables the pile to be well-aerated whilst acting as a visual screen for the first year's use.

The structure is surrounded by two closely planted rows of osier or biomass willow cuttings; this living wall of willow can then be woven into a hurdle-like structure and its annual growth can be harvested.

Spent Mushroom Compost

Spent mushroom substrate is the soil-like material remaining after a crop of mushrooms. Spent substrate is high in organic matter making it desirable for use as a soil amendment or soil conditioner.

Sometimes this material is called spent mushroom compost. This fact sheet briefly explains mushroom growing, so that the reader knows what is in the prepared substrate, and then describes the characteristics and possible uses of the material.

Mushroom Growing

Substrate prepared specifically for growing mushrooms is a blend of natural products. Common ingredients are wheat straw bedding containing horse manure, hay, corn cobs, cottonseed hulls, poultry manure, brewer's grain, cottonseed meal, cocoa bean hulls and gypsum. Growers may add ground soybeans or seed meal supplements later in the production cycle. On top of the substrate, farmers apply a "casing" layer, which is a mixture of peat moss and ground limestone. The casing material provides support for the growing mushrooms.

Spent mushroom substrate still has some nutrients available for the mushroom; however, it is more economical to replace the substrate and start a new crop. Before removing the spent substrate from the mushroom house, the grower "pasteurizes" it with steam to kill any pests or pathogens that may be present in the substrate and casing. This final pasteurization kills weed seeds, insects, and organisms that may cause mushroom diseases. Users may consider spent substrate clean of weed seeds and insects.

Mushroom growers sometimes apply a registered pesticide during the crop cycle. The local garden center sells most of the same pesticides a mushroom farmer uses. Even if pesticides have been applied, they are generally hard to find for two reasons. Organic matter in the substrate effectively binds pesticides. Also, these compounds decompose rapidly at the high temperatures used for pasteurizing the completed crop. It is safe to assume that the pesticide residue on spent substrate is low. Some farms are strictly "organic" and will not use chemical pesticides. These farms can be identified by contacting your Extension office.

Characteristics of Spent Mushroom Substrate

The typical composition of spent mushroom substrate fresh from a mushroom house will vary slightly. Since raw materials and other cultural practices change, each load of fresh spent substrate has a slightly different element and mineral analysis. Therefore the characteristics shown in Table indicate a range of values for each component. Sometimes, fresh substrate is placed in fields for at least one winter season and then marketed as "weathered" mushroom soil. This aged material has slightly different characteristics because the microbial activity in the field will change the composition and texture. The salt content may change during the aging period. If you have any specific questions concerning characteristics of either fresh or aged spent substrate, please contact your local Extension agent.

Appropriate uses of Spent Substrate

There are many appropriate uses for spent mushroom substrate. Spent mushroom substrate is excellent to spread on top of newly seeded lawns. The material provides cover against birds eating the seeds and will hold the water in the soil while the seeds germinate. Since some plants and garden vegetables are sensitive to high salt content in soils, avoid using fresh spent substrate around those plants. You may use spent substrate weathered for 6 months or longer in all gardens and with most plants. Obtaining spent substrate in the fall and winter, allowing it to weather, will make it ready to use in a garden the following spring. Spring and summer are the best time to use weathered material as a mulch.

As a soil amendment, spent substrate adds organic matter and structure to the soil. Spent substrate primarily improves soil structure and it does provide a few nutrients. Spent substrate is the choice ingredient by those companies making the potting mixtures sold in supermarkets or garden centers. These companies use spent substrate when they need a material to enhance the structure of a soil.

Table: Average analysis of spent mushroom substrate.

Contents	Units	Avg. Fresh	Weathered 16 mos.
Sodium, Na	% Dry Wt.	0.21 - 0.33	0.06
Potassium, K	% Dry Wt.	1.93 - 2.58	0.43
Magnesium, Mg	% Dry Wt.	0.45 - 0.82	0.88
Calcium, Ca	% Dry Wt.	3.63 - 5.15	6.27
Aluminum, Al	% Dry Wt.	0.17 -0.28	0.58
Iron, Fe	% Dry Wt.	0.18 - 0.34	0.58

Phosphorus, P	% Dry Wt.	0.45 - 0.69	0.84
Ammonia-N,NH4	% Dry Wt.	0.06 -0.24	0.00
Organic Nitrogen	% Dry Wt.	1.25 - 2.15	2.72
Total Nitrogen	% Dry Wt.	1.42 - 2.05	2.72
Solids	% Dry Wt.	33.07 - 40.26	53.47
Volatile Solids	% Dry Wt.	52.49 - 72.42	54.24
pH	Standard Units	5.8 - 7.7	7.1
N-P-K ratio	PPM Dry Wt.	1.8 - 0.6 - 2.2	2.7 – 0.8 – 0.47
% x 10,000 = PPM			

References

- Vertical-gardening: hydroponicsequipment.co, Retrieved 26 August, 2019

- Lorenz-Ladener, Hrsg. Claudia; Berger, Wolfgang (2005). Kompost-Toiletten: Wege zur sinnvollen Fäkalienentsorgung (1. überarb. u. erw. Aufl. ed.). Staufen im Breisgau: Ökobuch. p. 178. ISBN 978-3-936896-16-9

- Jana Voříškovái; Petr Baldrian (11 October 2012). "Fungal community on decomposing leaf litter undergoes rapid successional changes". The ISME Journal. 7 (3): 477–486. doi:10.1038/ismej.2012.116. PMC 3578564. PMID 23051693

- Mt. Pleasant, Jane (2006). "38". In John E. Staller; Robert H. Tykot; Bruce F. Benz (eds.). The science behind the Three Sisters mound system: An agronomic assessment of an indigenous agricultural system in the northeast. Histories of Maize: Multidisciplinary approaches to the prehistory, linguistics, biogeography, domestication, and evolution of maize. Amsterdam: Academic Press. pp. 529–537. ISBN 978-1-5987-4496-5

- Spent-mushroom-substrate: extension.psu.edu, Retrieved 24 March, 2019

Techniques used in Permaculture

5

- **Companion Planting**
- **Forest Gardening**
- **Grassed Waterway**
- **Holzer Permaculture**
- **Mycoforestry**
- **Intercropping**
- **Keyline Design**
- **Raised-bed Gardening**
- **Sheet Mulching**
- **Vegan Organic Gardening**
- **Companion Planting**
- **Waru Waru**

Companion planting, forest gardening, grassed waterway, holzer permaculture, mycoforestry, intercropping, keyline design, raised-bed gardening, sheet mulching, vegan organic gardening, companion planting, etc. are the example of various techniques used in permaculture. The aim of this chapter is to explore these techniques of permaculture.

Permaculture techniques are methods and practices that support the health of the overall system as well as individual structures you've put in place. There are a ton of possible permaculture techniques available for you to use; here are some of the most essential:

Companion Planting

Companion planting in gardening and agriculture is the planting of different crops in proximity for any of a number of different reasons, including pest control, pollination, providing habitat for beneficial insects, maximizing use of space, and to otherwise increase crop productivity. Companion planting is a form of polyculture.

Companion planting is used by farmers and gardeners in both industrialized and developing countries for many reasons. Many of the modern principles of companion planting were present many centuries ago in cottage gardens in England and forest gardens in Asia, and thousands of years ago in Mesoamerica.

Companion planting of carrots and onions.

Mechanisms

Companion planting can operate through a variety of mechanisms, which may sometimes be combined.

Provision of Nutrients

Legumes such as clover provide nitrogen compounds to other plants such as grasses by fixing nitrogen from the air with symbiotic bacteria in their root nodules.

Dandelions have long taproots that bring nutrients from deep within the soil to near the surface, benefitting neighboring plants that are shallower-rooted.

Trap Cropping

Trap cropping uses alternative plants to attract pests away from a main crop. For example, nasturtium is a food plant of some caterpillars which feed primarily on members of the cabbage family (brassicas); some gardeners claim that planting them around brassicas protects the food crops from damage, as eggs of the pests are preferentially laid on the nasturtium. However, while many trap crops have successfully diverted pests off of focal crops in small scale greenhouse, garden and field experiments, only a small portion of these plants have been shown to reduce pest damage at larger commercial scales.

Host-finding Disruption

Recent studies on host-plant finding have shown that flying pests are far less successful if their host-plants are surrounded by any other plant or even "decoy-plants" made of green plastic, cardboard, or any other green material.

The host-plant finding process occurs in phases:

- The first phase is stimulation by odours characteristic to the host-plant. This induces the insect to try to land on the plant it seeks. But insects avoid landing on brown (bare) soil. So if only the host-plant is present, the insects will quasi-systematically find it by simply landing on the only green thing around. This is called (from the point of view of the insect) "appropriate landing". When it does an "inappropriate landing", it flies off to any other nearby patch of green. It eventually leaves the area if there are too many 'inappropriate' landings. The second phase of host-plant finding is for the insect to make short flights from leaf to leaf to assess the plant's overall suitability. The number of leaf-to-leaf flights varies according to the insect species and to the host-plant stimulus received from each leaf. The insect must accumulate sufficient stimuli from the host-plant to lay eggs; so it must make a certain number of consecutive 'appropriate' landings. Hence if it makes an 'inappropriate landing', the assessment of that plant is negative, and the insect must start the process anew. Thus it was shown that clover used as a ground cover had the same disruptive effect on eight pest species from four different insect orders. An experiment showed that 36% of cabbage root flies laid eggs beside cabbages growing in bare soil (which resulted in no crop), compared to only 7% beside cabbages growing in clover (which allowed a good crop). Simple decoys made of green cardboard also disrupted appropriate landings just as well as did the live ground cover.

Pest Suppression

Some companion plants help prevent pest insects or pathogenic fungi from damaging the crop, through chemical means. For example, the smell of the foliage of marigolds is claimed to deter aphids from feeding on neighbouring plants.

Predator Recruitment

Companion plants that produce copious nectar or pollen in a vegetable garden (insectary plants) may help encourage higher populations of beneficial insects that control pests, as some beneficial predatory insects only consume pests in their larval form and are nectar or pollen feeders in their adult form. For instance, marigolds with simple flowers attract nectar-feeding adult hoverflies, the larvae of which are predators of aphids.

Protective Shelter

Shade-grown coffee plantation in Costa Rica. The red trees in the background provide shade; those in the foreground have been pruned to allow full exposure to the sun.

Some crops are grown under the protective shelter of different kinds of plant, whether as wind breaks or for shade. For example, shade-grown coffee, especially, has traditionally been grown in light shade created by scattered trees with a thin canopy, allowing light through to the coffee bushes but protecting them from overheating. Suitable Asian trees include (tton tong or dadap), (khae falang), (khi lek), (khao dao sang), and, a useful timber tree.

Systems

Systems in use or being trialled include:

- Square foot gardening attempts to protect plants from many normal gardening problems, such as weed infestation, by packing them as closely together as possible, which is facilitated by using companion plants, which can be closer together than normal.

- Forest gardening, where companion plants are intermingled to create an actual ecosystem, emulates the interaction of up to seven levels of plants in a forest or woodland.

- Organic gardening makes frequent use of companion planting, since many other means of fertilizing, weed reduction and pest control are forbidden.

Forest Gardening

Forest gardening is a low-maintenance, sustainable, plant-based food production and agroforestry system based on woodland ecosystems, incorporating fruit and nut trees, shrubs, herbs, vines and perennial vegetables which have yields directly useful to humans. Making use of companion planting, these can be intermixed to grow in a succession of layers to build a woodland habitat.

Forest gardening is a prehistoric method of securing food in tropical areas. In the 1980s, Robert Hart coined the term "forest gardening" after adapting the principles and applying them to temperate climates.

Robert Hart's forest garden in Shropshire.

Forest gardens are probably the world's oldest form of land use and most resilient agroecosystem. They originated in prehistoric times along jungle-clad river banks and in the wet foothills of monsoon regions. In the gradual process of families improving their immediate environment, useful tree and vine species were identified, protected and improved whilst undesirable species were eliminated. Eventually superior foreign species were selected and incorporated into the gardens.

Forest gardens are still common in the tropics and known by various names such as: in Kerala in South India, Nepal, Zambia, Zimbabwe and Tanzania; in Sri Lanka; the "family orchards" of Mexico. These are also called agroforests and, where the wood components are short-statured, the term shrub garden is employed. Forest gardens have been shown to be a significant source of income and food security for local populations.

Robert Hart adapted forest gardening for the United Kingdom's temperate climate during the 1980s. His theories were later developed by Martin Crawford from the Agroforestry Research Trust and various permaculturalists such as Graham Bell, Patrick Whitefield, Dave Jacke and Geoff Lawton.

In Temperate Climates

Robert Hart, forest gardening pioneer.

Hart began farming at Wenlock Edge in Shropshire with the intention of providing a healthy and therapeutic environment for himself and his brother Lacon. Starting as relatively conventional smallholders, Hart soon discovered that maintaining large annual vegetable beds, rearing livestock and taking care of an orchard were tasks beyond their strength. However, a small bed of perennial vegetables and herbs he planted was looking after itself with little intervention.

Following Hart's adoption of a raw vegan diet for health and personal reasons, he replaced his farm animals with plants. The three main products from a forest garden are fruit, nuts and green leafy vegetables. He created a model forest garden from a 0.12 acre (500 m) orchard on his farm and intended naming his gardening method or. Hart later dropped these terms once he became aware that and were already being used to describe similar systems in other parts of the world. He was inspired by the forest farming methods of Toyohiko Kagawa and James Sholto Douglas, and the productivity of the Keralan home gardens as Hart explains: "From the agroforestry point of view, perhaps the world's most advanced country is the Indian state of Kerala, which boasts no fewer than three and a half million forest gardens As an example of the extraordinary intensity of cultivation of some forest gardens, one plot of only 0.12 hectares (0.30 acres) was found by a study group to have twenty-three young coconut palms, twelve cloves, fifty-six bananas, and forty-nine pineapples, with thirty pepper vines trained up its trees. In addition, the small holder grew fodder for his house-cow".

Seven-layer System

The seven layers of the forest garden.

Robert Hart pioneered a system based on the observation that the natural forest can be divided into distinct levels. He used intercropping to develop an existing small orchard of apples and pears into an edible polyculture landscape consisting of the following layers:

- 'Canopy layer' consisting of the original mature fruit trees.

- 'Low-tree layer' of smaller nut and fruit trees on dwarfing root stocks.

- 'Shrub layer' of fruit bushes such as currants and berries.

- 'Herbaceous layer' of perennial vegetables and herbs.

- 'Rhizosphere' or 'underground' dimension of plants grown for their roots and tubers.

- 'Ground cover layer' of edible plants that spread horizontally.

- 'Vertical layer' of vines and climbers.

A key component of the seven-layer system was the plants he selected. Most of the traditional vegetable crops grown today, such as carrots, are sun-loving plants not well selected for the more shady forest garden system. Hart favoured shade-tolerant perennial vegetables.

Further Development

The Agroforestry Research Trust, managed by Martin Crawford, runs experimental forest gardening projects on a number of plots in Devon, United Kingdom. Crawford describes a forest garden as a low-maintenance way of sustainably producing food and other household products.

Ken Fern had the idea that for a successful temperate forest garden a wider range of edible shade tolerant plants would need to be used. To this end, Fern created the organisation Plants for a Future which compiled a plant database suitable for such a system. Fern used the term, rather than forest gardening, in his book.

Kathleen Jannaway, the founder of Movement for Compassionate Living (MCL), wrote a book outlining a sustainable vegan future called in 1991. The MCL promotes forest gardening and other types of vegan organic gardening. In 2009 it provided a grant of £1,000 to the Bangor Forest Garden project in Gwynedd, North West Wales.

Kevin Bradley in the US called his property and nursery "Edible Forest" in 1985, which combined trees and field crops. Today, his business and the 2005 book have spawned little "edible forests" all over the world.

Bill Mollison, who coined the term, visited Robert Hart at his forest garden in Wenlock Edge in October 1990. Hart's seven-layer system has since been adopted as a common permaculture design element.

Numerous permaculturalists are proponents of forest gardens, or food forests, such as Graham Bell, Patrick Whitefield, Dave Jacke, Eric Toensmeier and Geoff Lawton. Bell started building his forest garden in 1991 and wrote the book in 1995, Whitefield wrote the book in 2002, Jacke and Toensmeier co-authored the two volume book set in 2005, and Lawton presented the film in 2008.

In Tropical Climates

Forest gardens, or home gardens, are common in the tropics, using intercropping to cultivate trees, crops, and livestock on the same land. In Kerala in south India as well as in northeastern India, the home garden is the most common form of land use and is also found in Indonesia. One example combines coconut, black pepper, cocoa and pineapple. These gardens exemplify polyculture, and conserve much crop genetic diversity and heirloom plants that are not found in monocultures. Forest gardens have been loosely compared to the religious concept of the Garden of Eden.

Americas

The BBC's claimed that the Amazon rainforest, rather than being a pristine wilderness, has been shaped by humans for at least 11,000 years through practices such as forest gardening and. Since the 1970s, numerous geoglyphs have been discovered on deforested land in the Amazon rainforest, furthering the evidence of Pre-Columbian civilizations.

On the Yucatán Peninsula, much of the Maya food supply was grown in "orchard gardens", known as. The system takes its name from the low wall of stones (meaning 'circular' and, 'wall of loose stones') that characteristically surrounds the gardens.

Africa

In many African countries, for example Zambia, Zimbabwe, Ethiopia and Tanzania, gardens are widespread in rural, periurban, and urban areas and they play an essential role in establishing food security. Most well known are the Chaga or Chagga gardens on the slopes of Mt. Kilimanjaro in Tanzania. These are an example of an agroforestry system. In many countries, women are the main actors in home gardening and food is mainly produced for subsistence. In North Africa, oasis-layered gardening with palm trees, fruit trees, and vegetables is a traditional type of forest garden.

Projects

El Pilar on the Belize–Guatemala border features a forest garden to demonstrate traditional Maya agricultural practices. A further one acre model forest garden, called Känan K'aax (meaning 'well-tended garden' in Mayan), is funded by the National Geographic Society and developed at Santa Familia Primary School in Cayo.

In the United States, the largest known food forest on public land is believed to be the seven acre Beacon Food Forest in Seattle, Washington. Other forest garden projects include those at the central Rocky Mountain Permaculture Institute in Basalt, Colorado and Montview Neighborhood farm in Northampton, Massachusetts. The Boston Food Forest Coalition promotes local forest gardens.

In Canada Richard Walker has been developing and maintaining food forests in British Columbia for over 30 years. He developed a three acre food forest that at maturity provided raw materials for a plant nursery and herbal business as well as food for his family. The Living Centre has developed various forest garden projects in Ontario.

In the United Kingdom, other than those run by the Agroforestry Research Trust (ART), there are numerous forest garden projects such as the Bangor Forest Garden in Gwynedd, northwest Wales. Martin Crawford from ART administers the Forest Garden Network, an informal network of people and organisations who are cultivating forest gardens.

Grassed Waterway

A grassed waterway consists in a 2-metre (6.6 ft) to 48-metre-wide (157 ft) native grassland strip of green belt. It is generally installed in the thalweg, the deepest continuous line along a valley or watercourse, of a cultivated dry valley in order to control erosion. A study carried out on a grassed waterway during 8 years in Bavaria showed that it can lead to several other types of positive impacts, e.g. on biodiversity.

Grassed waterway in Velm, Belgium, during a sunny day.

Distinctions

Confusion between "grassed waterway" and "vegetative filter strips" should be avoided. The latter are generally narrower (only a few metres wide) and rather installed along rivers as well as along or within cultivated fields. However, buffer strip can be a synonym, with shrubs and trees added to the plant component, as does a riparian zone.

Runoff and Erosion Mitigation

Grassed waterway in Velm, Belgium, after a thunderstorm.

Runoff generated on cropland during storms or long winter rains concentrates in the thalweg where it can lead to rill or gully erosion.

Rills and gullies further concentrate runoff and speed up its transfer, which can worsen damage occurring downstream. This can result in a muddy flood.

In this context, a grassed waterway allows increasing soil cohesion and roughness. It also prevents the formation of rills and gullies. Furthermore, it can slow down runoff and allow its re-infiltration during long winter rains. In contrast, its infiltration capacity is generally not sufficient to reinfiltrate runoff produced by heavy spring and summer storms. It can therefore be useful to combine it with extra measures, like the installation of earthen dams across the grassed waterway, in order to buffer runoff temporarily.

Design of Grassed Waterways

The designs of the grassed waterways are similar to the design of the irrigation channels and are designed based on their functional requirements. Generally, these waterways are designed for carrying the maximum runoff for a 10- year recurrence interval period. The rational formula is invariably used to determine the peak runoff rate. Waterways can be shorter in length or sometimes, can be even very long. For shorter lengths, the estimated flow at the waterways outlets forms the design criterion, and for longer lengths, a variable capacity waterway is designed to account for the changing drainage areas.

Size of Waterway

The size of the waterway depends upon the expected runoff. A 10 year recurrence interval is used to calculate the maximum expected runoff to the waterway. As the catchment area of the waterway increases towards the outlet, the expected runoff is calculated for different reaches of the waterway and used for design purposes. The waterway is to be given greater cross-sectional area towards the outlet as the amount of water gradually increases towards the outlet. The cross-sectional area is calculated using the following formula:

$$a = \frac{Q}{V}$$

where,

- α = cross-sectional area of the channel,

- Q = expected maximum runoff,

- V = velocity of flow.

Shape of Water Way

The shape of the waterway depends upon the field conditions and type of the construction equipment used. The three common shapes adopted are trapezoidal, triangular, and parabolic shapes. In course of time due to flow of water and sediment depositions, the waterways assume an irregular shape nearing the parabolic shape. If the farm machinery has to cross the waterways, parabolic shape or trapezoidal shape with very flat side slopes are preferred. The geometric characteristics of different waterways are shown in figure for trapezoidal and parabolic waterways respectively.

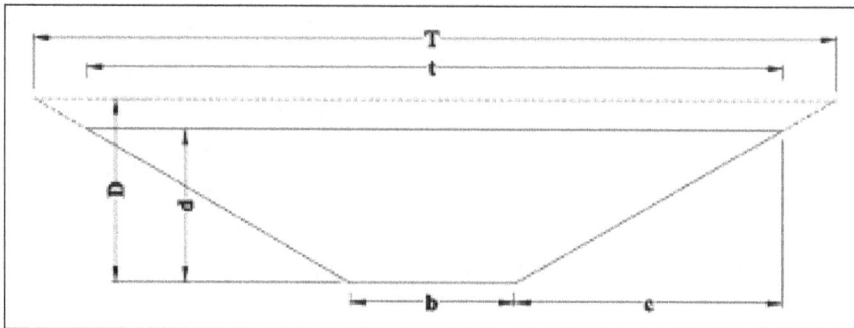

Trapezoidal Cross-section.

In the figure, d is the depth of water flow, b is bottom width, t is the top width of maximum water conveyance, T is top width after considering free board depth, (D - d) is the free board and slope (z) is c/d.

Table: Design Dimensions for Trapezoidal Cross-section.

Cross-sectional Aria, a	Wetted perimeter, P	Hydraulic Radius, $R = \dfrac{a}{p}$	Top width
$bd + zd^2$ Where, Z=c/d	$b + 2d\sqrt{Z^2 + 1}$	$\dfrac{bd + zd^2}{b + 2d\sqrt{z^2 + 1}}$	$T = b + 2dz$ $T = b + 2Dz$

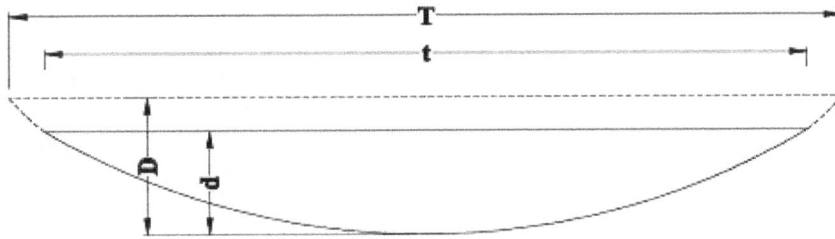

Parabolic Cross-section.

Table: Design Dimensions for Parabolic Cross-Section.

Cross-sectional Aria, a	Wetted perimeter, P	Hydraulic Radius, $R = \dfrac{a}{p}$	Top width
$\dfrac{2}{3}td$	$t + \dfrac{8d^2}{3t}$	$\dfrac{t^2 \times d}{1.5t^2 + 4d^2}$ $\dfrac{2d}{3} approx$	$t = \dfrac{a}{0.6_7 d}$ $T = t\left(\dfrac{D}{d}\right)^{\frac{1}{2}}$

Channel Flow Velocity

The velocity of flow in a grassed waterway is dependent on the condition of the vegetation and the soil erodibility. It is recommended to have a uniform cover of vegetation over the channel surface to ensure channel stability and smooth flow. The velocity of flow through the grassed waterway depends upon the ability of the vegetation in the channel to resist erosion. Even though different types of grasses have different capabilities to resist erosion; an average of 1.0 m/sec to 2.5 m/sec are the average velocities used for design purposes. It may be noted that the average velocity of flow is higher than the actual velocity in contact with the bed of the channel. Velocity distribution in a grassed lined channel is shown in figure. Recommended velocities of flow based on the type of vegetation are shown in Table. The permissible velocities of flow on different types of soils are given in table.

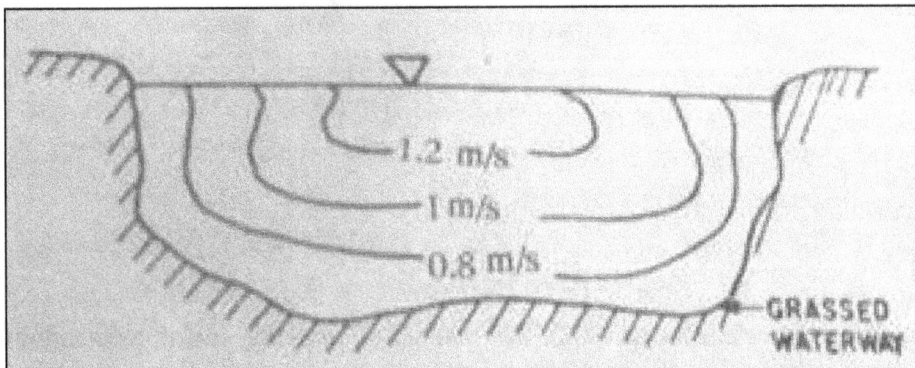

Velocity Distribution in Open Channel.

Table: Recommend Velocities of Flow in a Vegetated Channel.

Type of vegetation cover	Flow velocity, (m/s)	
	Type	Magnitude
Spare green cover	Low velocity	1-1.15
Good quality cover	Medium velocity	1.5-1.8
Excellent quality cover	High velocity	1.8-2.5

Table: Permissible Velocity of Flow on Different Types of Soil.

Type of soil	Permissible velocity, (m/s)	
	Clean water	Colloidal water
Very fine sand	0.45	0.75
Sandy loam	0.55	0.75
Silty loam	0.60	0.90
Alluvial silt without colloids	0.60	1.00
Dense clay	0.75	1.00
Hard clay, colloidal	1.10	1.50
Very hard clay	1.80	1.80
Fine gravel	0.75	1.50
Medium and coarse gravel	1.20	1.80
Stones	1.50	1.80

Design of Cross-section

The design of the cross-section is done using Equation for finding the area required and Manning's formula is used for cross checking the velocity. A trial procedure is adopted. For required cross-sectional area, the dimensions of the channel section are assumed. Using hydraulic property of the assumed section, the average velocity of flow through the channel cross-section is calculated using the Manning's formula as below:

$$V = \frac{S^{1/2} R^{2/3}}{n}$$

where,

- V = velocity of flow in m/s;
- S = energy slope in m/m;
- R = hydraulic mean radius of the section in m;
- n = Manning's roughness coefficient.

The Manning's roughness coefficient is to be selected depending on the existing and proposed vegetation to be established in the bed of the channel. Velocity is not an independent parameter. It will depend on n which is already fixed according to vegetation, R which is a function of the channel geometry and slope S for uniform flow. Slope S has to be adjusted. If the existing land slope gives high velocity, alignment of the channel has to be changed to get the desired velocity.

Problem: Design a grassed waterway of parabolic shape to carry a flow of 2.6 m/s down a slope of 3 percent. The waterway has a good stand of grass and a velocity of 1.75 m/s can be allowed. Assume the value of in Manning's formula as 0.04.

Solution: Using, Q = AV for a velocity of 1.75 m/s, a cross-section of 2.6/1.75 = 1.485 m (~1.5 m) is needed.

Assuming, t = 4 m, d = 60 cm.

$$A = \frac{2}{3}t \times d = \frac{2}{3}4 \times 0.6 = 1.6 m^2$$

$$P = t + 8\frac{d^2}{3t} = 4 + 8\frac{(0.6)^2}{3 \times 4} = 4.24m$$

$$R = \frac{A}{P} = \frac{1.6}{4.24} = 0.377m$$

$$V = \frac{S^{1/2}R^{2/3}}{n} = \frac{(0.03)^{1/2} \times (0.377)^{2/3}}{0.04} 2.26m/s$$

The velocity exceeds the permissible limit. Assuming a revised

t = 6 m and d = 0.4 m.

$$A = \frac{2}{3}t \times d = \frac{2}{3}4 \times 0.6 = 1.6 m^2$$

$$P = t + 8\frac{d^2}{3t} = 6 + 8\frac{(0.4)^2}{3 \times 6} = 6.45m$$

$$V = \frac{S^{1/2}R^{2/3}}{n} = \frac{(0.03)^{1/2} \times (0.248)^{2/3}}{0.04} = 1.70m/s$$

The velocity is within the permissible limit.

Q = 1.6 × 1.7 = 2.72 m/s

The carrying capacity (Q) of the waterway is more than the required. Hence, the design of water-way is satisfactory. A suitable freeboard to the depth is to be provided in the final dimensions.

Construction of the Waterways

It is advantageous to construct the waterways at least one season before the bunding. It will give time for the grasses to get established in the waterways. First, unnecessary vegetation like shrubs etc. are removed from the area is marked for the waterways. The area is then ploughed if necessary and smoothened. Establishment of the grass is done either by seeding or sodding technique. Maintenance of the waterways is important for their proper operation. Removal of weeds, filling of the patches with grass and proper cutting of the grass are of the common maintenance operations that should be followed for an efficient use of waterways.

Selection of Suitable Grasses

The soil and climate conditions are the primary factors in selection of vegetations to be established for construction of grassed waterways. The other factors to be considered for selection of suitable grasses are duration of establishment, volume and velocity of runoff, ease of establishment and time required to develop a good vegetative cover. Furthermore, the suitability of the vegetation for utilization as feed or hay, spreading of vegetation to the adjoining fields, cost and availability of seeds and redundancy to shallow flows in relation to the sedimentation are the important factors that should be considered for the selection of vegetation.

Generally, the rhizomatous grasses are preferred for the waterway, because they get spread very quickly and provide more protection to the channel than the brush grasses. Deep rooted legumes are seldom used for grassed waterways, because they have the tendency to loosen the soil and thus make the soil more erodible under the effect of fast flowing runoff water. Sometimes, a light seeding of small grain is also used to develop a quick cover before the grasses are fully established in the waterway.

Construction Procedure and Maintenance

Ordinary tools such as slip scraper can be easily used for construction of waterways. However, the use of grader blade or a bulldozer can be preferred, particularly when a considerable earth movement is needed. Since the channel is prone to erosion before vegetations are established, it is very essential to construct the waterway when the field is in meadow and the amount of runoff from the area is also very less. In addition, if the erosion hazard is very high, then runoff should also be essentially diverted from the waterway until a good grassed cover is developed in the waterway.

The construction of grassed waterways is carried out using the following steps.

Step-1: Shaping

The shaping of the waterway should be done as straight and even as possible. Any sudden fall or sharp turn must be eliminated, except in the area where the structure is planned to be installed in the waterway. In addition, the grade should also be shaped according to the designed plan. Also, the stones and stumps which are likely to interfere with the discharge rate must be removed.

Step-2: Grass Planting

After shaping the waterway channel, the planting of grasses is very important. Priorities should always be given to the local species of grasses. The short forming or rhizome grasses are more preferable as compared to the tall bunch type grasses.

In large waterways, the seeding is cheaper than the sodding. Therefore, the seeding should be preferred for grass development. It is also suggested that the seeded area should be mulched especially for production purposes. Immediately after grass planting, the waterways should not be allowed for runoff flow.

Step 3: Ballasting

Ballasting is done in those localities where rocks are readily available adjacent to the sites and waterway gradient is very steep. Ballasting is generally recommended for the waterways in the small farms. The stones to be used for this purpose should be at least of 15 to 20 cm diameter; and they should be placed firmly on the ground. From stability point of view, on very steep slopes, wire mesh should be used to encase the stones. In parabolic shaped waterways, partial ballasting should be done in the centre, leaving the sides with grass protection.

Step 4: Placing of Structure

Structures (drop) are essential if there is sudden fall in the waterway flow path. Because under this situation, there is a possibility of soil scouring due to falling of water flow from a higher elevation to a lower elevation. For eliminating this problem, the constructed structure must be sufficiently strong to handle the designed flows successfully. As a precautionary measure, care should be taken to see that the water must not flow from the below or around the structure but through the top of the structure. In addition, the structure should be constructed on firm soils with strong and deep foundation. The apron or stilling basin of drop structures should be sufficiently strong and able to absorb or dissipate the energy/impact of falling water. After construction, earth filling should be done around the structure and it should be properly consolidated to prevent further settlement. Proper sodding should also be provided at the junction of earth filling and the structure to prevent tunneling.

Maintenance

The grasses grown in waterway should always be kept short and flexible, so that they shingle as water flows over them, but do not lodge permanently. For this purpose, the grass should be mowed two to three times in a year. The mowed grasses must be removed from the waterway, so that they do not get accumulated at some spots in the waterway and also should not obstruct the flow. The deposition of mowed grasses in the section of the waterway reduces the flow capacity of the waterway and also diverts the direction of flowing water which can cause turbulence and thus damage of the channel. It is also possible to keep the grasses short by light pasturing, which should not be done in wet condition. When the grass is pastured, it is necessary to apply manure to discourage grazing. The waterway should not be used as a road for livestock. After the vegetative cover is established and runoff passes through them for a long time, a light application of fertilizer should be done because the flowing runoff removes the plant food from the soil of waterway.

Similarly, if waterways are to be crossed by tillage implements, they should be disengaged, plough should be lifted and disc straightened. Tillage operation should also be done following nearly the contour. The waterway and its sides should not be touched during tillage operation. It is also essential that if there is any damage of the waterway, it should be quickly repaired so that the damage may not enlarge due to rainfalls. Overall, it should always be remembered that the waterways are an integral part of watershed conservation or land treatment system. If they fail to handle the peak discharge due to lack of proper maintenance, then the prolong flow of runoff through them can develop gullies in the area. Briefly, the maintenance of waterways can be taken up using the following process.

- The outlets should be safe and open so as not to impede the free flow.

- Grassed waterways should not be used as footpaths, animal tracks, or as grazing grounds.

- Frequent crossing of waterways by wheeled vehicles should not be allowed.

- Newly established waterways should be kept under strict watch.

- The large waterways should be kept under protection with fencing.

- Waterways must be inspected frequently during first two rainy seasons, after construction.

- If there is any break in the channel or structures, then they should be repaired immediately.

- The bushes or large plants grown in the waterway should be removed immediately as they may endanger the growth of grasses.

- The level of grass in waterway should be kept as low and uniform as possible to avoid turbulent flow.

Holzer Permaculture

The Holzer Permaculture is a branch of permaculture developed independently from the mainstream permaculture in Austria by Sepp Holzer. It is particularly noteworthy because it grew out of practical application and was relatively detached from the scientific community.

Sepp Holzer started reorganising his father's property according to ecological patterns in the early 1960s after he took over the farm. As an adolescent he conducted layman experiments with plants native to the area and learned from his own observations.

Since having taken over his father's property, he has expanded it from 24 to 45 hectares. according to his methods together with his wife.

His expanded farm now spans over 45 hectares of forest gardens, including 70 ponds, and is said to be the most consistent example of permaculture worldwide. In the past he has experimented with many different animals. As a result of these experiments, there is a huge role for animals in the Holzer Permaculture.

He has created some of the world's best examples of using ponds as reflectors to increase solar gain for passive solar heating of structures, and of using the microclimate created by rock outcrops to effectively change the hardiness zone for nearby plants. He has also done original work in the use of Hügelkultur and natural branch development instead of pruning to allow fruit trees to survive high altitudes and harsh winters.

Comparison to Regular Permaculture

It is difficult to make out differences between the methods and practices of Sepp Holzer in contrast to the more scientific and theoretical permacultural mainstream. Nevertheless, here are some major points to consider:

- His designs are mostly aimed at raising temperatures and creating micro-climates with rocks, ponds and living wind barriers, in an area with 4 °C on the average and -20 °C in the winter. The ponds he makes do not contain any pond liner. Instead, he makes the ponds watertight by sifting the fine from the coarse soil in the earth pile dug up with the excavator. The excavator is then used to pile only the coarse soil up into walls which are then tampered down using the excavator's bucket. The bottom of the pond, he makes watertight by vibrating the excavator's bucket when the pond has been filled with 30–40 cm of water.

- Another aspect was the necessity of creating terraces on his farm's hillsides leading him to the use of heavy machinery (like excavators). Many of the terraces he construct are also given a humus storage ditch (placed in between the terraces).

- Famous is his Hügelkultur technique, which is basically the use of raised beds in which he uses bulky material such as tree trunks. On his farm, he made a pick-your-own area where visitors can pick their own produce from the raised beds and then pay for it at a counter upon leaving the area.

- He uses animal labor alongside human labor, working his farm with only two people. He optimizes the natural patterns of animal behavior to reduce human or machine-driven labor. As an example: he uses swine to "plow" new beds for sowing. This is a very effective way of digging, as the only thing he has to do is to throw some corn and fruit on the spot he wants dug up. A couple of days later, he can bring the pigs back to their enclosure and plant new plants in the bed. Holzer is able to successfully grow his plants without using any fertilizer. The animals he uses are all heirloom races, which are hardy and require (almost) no maintenance. Examples of the races of swine he uses are Mangalitza, Swabian Hall swine, Duroc, Turopolje. He also keeps quail, capercaillie, hazel grouse, wisent, scottish highland cattle, hungarian steppe cattle, Dahomey miniature cattle, American bison, yak, water buffalo. He has the animals live outside, in paddocks/shelters and has them share space with orchards and forests. Many of the fruit trees in the orchards (especially those on very sloped terrain) are used exclusively to feed the animals. They collect it as it falls to the ground (those trees are hence not being picked/nor is produce of it sold).

- He also does not prune fruit trees much nor does he cut the lower branches on fruit trees (as this can hurt the tree -due to the lignification not being able to complete before frost and the fact that the unpruned fruit trees survives snow loads that will break pruned trees. He also says that leaving the branches on protects the tree against browsing by animals. He does however makes a point of using deep-rooted pioneer plants such as lupins, sweet clover, lucerne and broom. These crops are said to aerate the soil and make sure no water is left standing near the tree. No use is made of wire meshes as protection against voles, since he states they are not efficient in preventing damage from voles using this technique anyway. He makes no use of contemporary fruit tree cultivars, but only uses (very strong/hardy) old, local (heirloom) cultivars. In addition to (old cultivars of) regular fruit trees, he also plants much fruit tree species that are specific to use in phytotherapy or which can only function as animal feed (i.e. crab apple, wild pear, wild cherry, blackthorn, rowan, wild service, service tree, cornelian cherry, snowy mespilus). Similar to mainstream permaculture, he makes no use of chemical fertiliser or pesticides, at all. Due to the altitude he's at, his trees bear fruit later, meaning he can sell it after most (sealevel) farmers sold their produce. Due to this (and the fact that he produces heirloom fruit varieties, and not regular varieties), he is often able to get a better price for it. In some cases, customers (like distilleries) are even willing to pick the fruit themselves, eliminating the labour expense for him.

- He grows many old cereals on his farm, such as einkorn, emmer, black emmer, spelt, fichtelgebirgshafer, wild rye, black oats, naked oats, barley, Siberian grain (secale cereale, russian cultivar), tauernroggen.

- He also makes much use of green manure crops (like stinging nettle, phacelia, yellow, white and narrow-leaved lupin, garden pea, grass pea, fodder & Kidney vetch, yellow, subterranean, Crimson, Persian, Egyptian, red & white sweet clover, Birdsfoot trefoil, lucerne, black medick, Sainfoin, Serradella, fiddleneck, sunflowers, Jerusalem artichoke, Gold-of-pleasure) and grows these crops extensively on his farm. He leaves them standing in autumn, rather than digging them in. He instead relies on natural decay of the plants. He often relies on the natural spreading of the seeds of the crops for their re-sowing.

- Another aspect is the abandonment of other horticultural principles such as intercropping plants with very high and very low pH requirements (for example, Rhododendron with roses). Instead, Holzer mixes thirty or more different types of seeds in a bucket and tosses the mix richly onto a larger area.

The Krameterhof

Situated in Ramingstein on the slopes of Mount Schwarzenberg his farm (Krameterhof) lies at varying elevations ranging from 1100 to 1500 metres above sea level.

The exceptionally harsh climatic conditions in the area are generally considered inappropriate for farming. Nevertheless, he has managed to cultivate a variety of crops and even exotic plants like Kiwis and Sweet Chestnut.

The Krameterhof is less an operational enterprise, in terms of crop-yield (although it does provide numerous sorts of produce for the community), and more a fully functional showcase or research station for permaculture.

Endangered livestock species and rare alpine- and cultural plant species are integrated into the farm.

Mycoforestry

Mycoforestry is an ecological forest management system implemented to enhance forest ecosystems and plant communities through the introduction of mycorrhizal and saprotrophic fungi.

Mycoforestry is considered a type of permaculture and can be implemented as a beneficial component of an agroforestry system. Mycoforestry can enhance the yields of tree crops and produce edible mushrooms, an economically valuable product. By integrating plant-fungal associations into a forestry management system, native forests can be preserved, wood waste can be recycled back into the ecosystem, planted restoration sites are enhanced, and the sustainability of forest ecosystems are improved. Mycoforestry is an alternative to the practice of clearcutting, which removes dead wood from forests, thereby diminishing nutrient availability and reducing soil depth.

Species are ectomycorrhizal with many trees.

Selection of Fungal Species

According to Paul Stamets, the first principle for the creation of a mycoforestry system is to utilize native fungal species. Implementing a mycoforestry system provides the potential of improving restoration efforts and the possibility of economic gain through mushroom cropping and harvesting. However to utilize native fungal flora, first the relationships between present fungal species and growth substrate, and habitat need to be studied.

A simple way to introduce a mycoforestry system and enhance out-plantings for crops and forest restoration sites is to "use mycorrhizal spore inoculum when replanting forest lands" For this process it is best to match native trees with native mycorrhizal fungi. This method keeps and will promote the functioning of the native ecosystem, and native biodiversity.

It is assumed in a functioning forest ecosystem an underground mycelial network persists even if no fruiting bodies are visible. A period of disappearance of mushrooms from an area should not cause alarm. In order to trigger the formation of fruiting bodies, many fungal species require specific environmental conditions. Most species of fungi do not fruit year round.

Mycoforestry is an emergent scientific field and practice. Until broadly standardized protocols are created and perfected, the collection of both current and historical ecological site conditions will improve the success of the project. Therefore, a survey of fungal relations at the site under both prime and poor conditions is beneficial to implementation of a mycoforestry system.

Saprotrophic Fungi

The second principle is to promote saprotrophic fungi in the environment. Saprophytic fungi are crucial to mycoforestry systems because these are the primary composers breaking down wood and returning nutrients to the soil for use by the rest of the forest ecosystem. This can be accomplished through inoculation of wood debris at site. Spored oils can be used in chain-saws when problematic or invasive hardwood requires felling. This method is a simple means to inoculate a tree. Additionally plug spawn can be implemented and injected into wood mass again prompting colonization by the selected fungus. Eventually repeated colonization efforts should not be necessary as many fungal life forms are strong and will spread and sustain in the soil on their own.

Edible oyster mushrooms (sp.) fruiting from a stump.

In management of the mycoforestry system it is important that dead wood be in contact with the ground. This allows fungus to reach up from the soil and decompose fallen wood releasing nutrients at a much quicker rate then if the wood is left standing. Additionally it is important to leave dead wood on site for decomposition back into the soil. This philosophy is similarly based to the fact that clear cutting of a forest reduces soil nutrients and thickness.

Beneficial Fungal Interactions

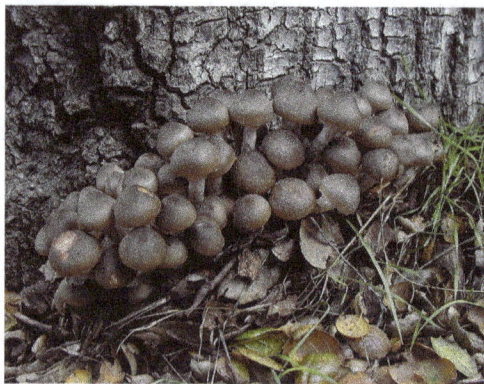
A parasitic fungus.

The third principal is to implement species known to benefit plant species. These are commonly mycorrhizal fungus that form long term associations with plants, often extending inside of plants roots acting as an additional root system providing for better absorption of nutrients and water.

Utilizing mushroom species that attract insects could be a useful source of fish food. This practice makes the mycoforestry a larger system. Unlike most agriculture systems it helps the environment in a number of ways. It ties all biological aspects of the environment together, creating sustainable living and food production as well as sustainable fisheries similar to the ancient Hawaiian Ahupua'a, which utilized sustainable all portions of the land for environmental and food security.

Additionally fungal species can be implemented that compete with disease causing agents like Armillaria root rots to provide long term protection of the forestry system.

Additionally, the implementation of an agroforestry system performs mycoremediation and mycofiltration activities cleaning up toxins and restoring the environment.

Intercropping

Intercropping is a farming method that involves planting or growing more than one crop at the same time and on the same piece of land. It means having more than one type of crop growing in the same space at the same time. The rationale behind this farming practice is that different crops planted are not likely to share insects and disease-causing agents while the goal is to produce even greater yield than would be if space was utilized by one crop. However, the careless congregation of plants is not considered as intercropping.

Types of Intercropping

There are four types of Inter-cropping:

- Row Inter Cropping: Growing two or more crops at the same time with at least one crop planted in rows. Example: Annual crops such as rice, corn and pineapple are commonly grown as intercrop between the rows. Banana, Papaya, Coffee and Cacao are commonly grown in multiple rows.

- Strip Inter Cropping: In this method two or more crops grow together in strips wide enough to allow separate production of crops using mechanical implements but close enough for the crops to interact. For Example: Wheat, Soybean and Corn.

- Mixed Inter Cropping: Growing two or more crops at the same time with no distinct row arrangement. For Example: Annual crops such as growing bean, corn and squash.

- Relay Inter Cropping: Growing a second crop is planted into an existing crop when it has flowered but before harvesting. For Example: Cassava, Cotton and Sweet Potato.

Advantages and Disadvantages of Intercropping

Advantages of Intercropping

- Greater Income, Greater Yield: Intercropping offers greater financial returns for a farmer. Even if you are growing some produce for your own family or just as part of a hobby, you will have multiple types of produce, which is always a nice outcome. Intercropping will help farmers use the same land as available and yield more as well as diverse produce. This generates more income for the farmer without really taking up any major expenditure. The infrastructure available or the land used remains the same.

- Insurance against Crop Damage: Intercropping can be the insurance that farmers need, especially when the region is vulnerable to weather extremes. Drought, torrential rain, hurricanes or cyclones and various other weather elements can affect the yield of a given year or season. Having diverse yields allows the farmer to have some income even if the primary crop gets damaged or doesn't yield as much as expected.

- Optimum use of Soil: Intercropping makes the most of the available soil. When anything is grown on a farmland, the crop tends to absorb as much water and nutrients as it needs. There could be more nutrients in the soil under the crops and around. This soil and more specifically the nutrients can be used, by the different varieties of crops. Intercropping also averts soil runoff and can prevent the growth of weeds.

- Good for Primary Crops: Intercropping is good for the primary crops. The secondary crops can provide shelter and even protect the primary crops. Intercropping also allows you to grow cash crops or any crop that will actually supplement the primary crop in some way.

Disadvantages of Intercropping

- Poor Yields: Intercropping can lead to poorer yields. The crops may not be compatible. The crops may actually compete for the same nourishment and of course water, which may lead to an unmanageable conflict. It is possible that both crops don't yield enough produce.

- Costly and Complicated: Intercropping obviously costs more money upfront. There is a need for more fertilizers and water. Harvesting is also more complicated. If something goes horribly wrong with either crop then the other crop may also get damaged.

Keyline Design

Keyline design is a landscaping technique of maximizing the beneficial use of the water resources of a tract of land. The "keyline" denominates a specific topographic feature related to the natural flow of water on the tract. Keyline design is a system of principles and techniques of developing rural and urban landscapes to optimize use of their water resources.

A keyline irrigation channel in Orana, Australia.

Application

Keyline designs include irrigation dams equipped with through-the-wall lockpipe systems, gravity feed irrigation, stock water, and yard water. Graded earthen channels may be interlinked to broaden the catchment areas of high dams, conserve the height of water, and transfer rainfall runoff into the most efficient high dam sites. Roads follow both ridge lines and water channels to provide easier movement across the land.

Keyline Scale of Permanence

The foundation of Yeomans' Keyline design system is the Keyline Scale of Permanence (KSOP), which was the outcome of 15 years of adaptive experimentation. The Scale identifies the environmental elements of typical farms and orders them according to their degree of permanence as follows:

- Climate.

- Landshape (topography).

- Water supply.

- Roads and means of access.

- Trees.

- Structures (edifices).

- Subdivisional fences.

- Soil.

Keyline design considers these elements in planning the placement of water storage features, roads, trees, edifices, and fences. On undulating land, keyline design involves identifying ridges, valleys, and natural water courses and designing with them in mind in order to optimize water storage sites. Constructing interconnecting channels may be part of such optimization.

Rancho San Ricardo, Oaxaca, México.

The identified natural water lines delineate the possible locations for the various less permanent elements, e. g. roads, fences, trees, and edifices, which if so located would help optimize the natural potential of the land in question.

Keypoint

In a smooth, grassy valley, a location denominated the keypoint is identified at which the lower and leveller portion of the primary valley floor suddenly steepens higher. The keyline of this primary valley is determined by pegging a contour line that conforms to the natural shape of the valley through the keypoint, such that all points on the keyline are at the same elevation as the keypoint. Contour plowing both above and below the keyline and parallel to it ipso facto is "off-contour", but the developing pattern tends to drift rainwater runoff away from the center of the valley and, incidentally, prevent erosion of its soil.

Cultivation conforming to Keyline design for ridges is done parallel to any suitable contour, but only on the high side of the contour's guide line. This ipso facto develops a pattern of off-contour cultivation in which all the rip marks made in the soil slope down towards the center of the ridge. This pattern of cultivation allows more time for water to infiltrate. Cultivation following Keyline pattern also enables controlled flood irrigation of undulating land, which increases the rate of development of deep, fertile soil.

In many nations, including Australia, it is important to optimize infiltration of rainfall, and Keyline cultivation accomplishes this while delaying the concentration of runoff that could damage the land. Yeomans' technique differs from traditional contour plowing in several important respects. Random contour plowing also becomes off contour but usually with the opposite effect on runoff, namely causing it to quickly run off ridges and concentrate in valleys. The limitations of the traditional system of soil conservation, with its "safe disposal" approach to farm water, was an important motivation to develop Keyline design.

Applications

David Holmgren, one of the founders of permaculture, used Yeoman's Keyline design extensively in the formulation of his principles of permaculture and the design of sustainable human settlements and organic farms.

Darren J. Doherty has extensive global experience in Keyline design, development, management, and education. He uses Keyline as the basis for his Regrarians framework, which he considers a revision and synthesis of Keyline design, permaculture, holistic management, and several other innovative, human ecological frameworks into a coherent process-based system of design and management of regenerative economies.

A topographical example of Keyline design is available at (37°09′33″S 144°15′08″E37.159154°S 144.252248°E).

Keyline design also includes principles of rapidly enhancing soil fertility. Yeomans and his sons were also instrumental in the design and production of special plows and other equipment for Keyline cultivation.

Raised-bed Gardening

Raised-bed gardening is a form of gardening in which the soil is enclosed in three-to-four-foot-wide (1.0–1.2 m) containment units ("beds"), which are usually made of wood, rock, or concrete and which can be of any length or shape. The soil is raised above the surrounding soil (approximately six inches to waist-high) and may be enriched with compost.

Vegetable plants are spaced in geometric patterns, much closer together than in conventional row gardening. The spacing is such that when the vegetables are fully grown, their leaves just barely touch each other, creating a microclimate in which weed growth is suppressed and moisture is conserved.

Raised bed gardening.

Raised beds lend themselves to the development of complex agriculture systems that utilize many of the principles and methods of permaculture. They can be used effectively to control erosion and recycle and conserve water and nutrients by building them along contour lines on slopes. This also makes more space available for intensive crop production. They can be created over large areas with the use of several commonly available tractor-drawn implements and efficiently maintained, planted and harvested using hand tools.

This form of gardening is compatible with square foot gardening and companion planting.

Circular raised beds with a path to the center (a slice of the circle cut out) are called keyhole gardens. Often the center has a chimney of sorts built with sticks and then lined with feedbags or grasses that allows water placed at the center to flow out into the soil and reach the plants' roots.

A self watering raised bed known as a wicking bed is particularly beneficial in dry climates and are often made by converting Intermediate bulk container (IBC's).

Materials and Construction

Vegetable garden bed construction materials should be chosen carefully. Some concerns exist regarding the use of pressure-treated timber. Pine that was treated using chromated copper arsenate or CCA, a toxic chemical mix for preserving timber that may leach chemicals into the soil which in turn can be drawn up into the plants, is a concern for vegetable growers, where part or all of the plant is eaten. If using timber to raise the garden bed, ensure that it is an untreated hardwood to prevent the risk of chemicals leaching into the soil. A common approach is to use timber sleepers joined with steel rods to hold them together. Another approach is to use concrete blocks, although less aesthetically pleasing, they are inexpensive to source and easy to use.

On the market are also prefab raised garden bed solutions which are made from long lasting polyethylene that is UV stabilized and food grade so it will not leach undesirable chemicals into the soil or deteriorate in the elements. A double skinned wall provides an air pocket of insulation that minimizes the temperature fluctuations and drying out of the soil in the garden bed. Sometimes raised bed gardens are covered with clear plastic to protect the crops from wind and strong rains. Pre-manufactured raised bed gardening boxes also exist. There are variants of wood, metal, stone and plastic. Each material type has advantages and disadvantages.

Benefits

Raised beds produce a variety of benefits: they extend the planting season, they can reduce weeds if designed and planted properly, and they reduce the need to use poor native soil. Since the gardener does not walk on the raised beds, the soil is not compacted and the roots have an easier time growing. Waist-high raised beds enable the elderly and physically disabled to grow vegetables without having to bend over to tend them.

Sheet Mulching

In permaculture, sheet mulching is an agricultural no-dig gardening technique that attempts to mimic the natural soil-building process in forests. When deployed properly and in combination with other permaculture principles, it can generate healthy, productive, and low maintenance ecosystems.

Sheet mulching, also known as composting in place, mimics nature by breaking down organic material from the topmost layers down. The simplest form of sheet mulching consists of applying a bottom layer of decomposable material, such as cardboard or newspapers, to the ground to kill existing vegetation and suppress weeds. Then, a top layer of organic mulch is applied. More elaborate sheet mulching involves more layers. Sheet mulching is used to transform a variety of surfaces into a fertile soil that can be planted. Sheet mulching can be applied to a lawn, a dirt lot full of perennial weeds, an area with poor soil, or even pavement or a rooftop.

Technique

A model for sheet mulching consists of the following steps:

- The area of interest is flattened by trimming down existing plant species such as grasses.

- The soil is analyzed and its pH is adjusted (if needed).

- The soil is moisturized (if needed) to facilitate the activity of decomposers.

- The soil is then covered with a thin layer of slowly decomposing material (known as the weed barrier), typically cardboard. This suppresses the weeds by blocking sunlight, adds nutrients to the soil as weed matter quickly decays beneath the barrier, and increases the mechanical stability of the growing medium.

- A layer (around 10 cm thick) of weed-free soil, rich in nutrients is added, in an attempt to mimic the surface soil, or A horizon.

- A layer (at most 15 cm thick) of weed-free, woody and leafy matter is added in an attempt to mimic the forest floor, or O horizon. Theoretically, the soil is now ready to receive the desirable plant seeds or transplants, Variations and considerations.

- Often the barrier is applied a few months before planting to ensure the penetration of roots of newly planted seeds. Very thick barriers can cause anaerobic conditions.

- Some permaculturists incorporate composting in steps 5 and 6. Sheets of newspaper and clothing can be used instead of cardboard. Before step 4, an initial layer (2–3 kg/m) of matter rich in nutrients (such as compost or manure) may be added to bolster decomposition. Some varieties of grasses and weeds may be beneficial in a number of ways. Such plants can be controlled and used rather than eradicated.

- One variation of mulching, called hugelkultur, involves using buried logs and branches as the first layer of the bed.

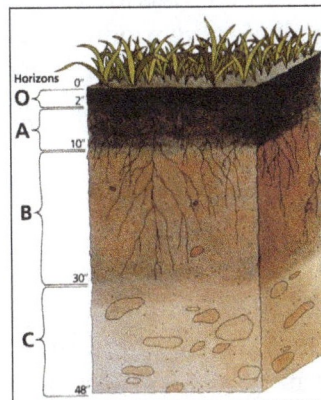

Typical layers of natural soil.

Advantages

Sheet mulch has important advantages relative to conventional methods, such as tilling, plowing or applying herbicides:

- Improvement of desirable plants' health and productivity. Retention of water and nutrients and stabilization of biochemical cycles. Improvement of soil structure, soil life, and prevention of soil erosion. Avoidance of potentially dangerous pesticides, especially herbicides.

- Reduction of overall maintenance labor and costs. Most of the materials required to sheet mulch can be collected at no cost, and materials can be substituted for those readily available in certain areas. For instance, suburban areas may have a plentiful supply of leaves, and farming communities may have spoiled hay and manure.

Disadvantages

- Some weed seeds (such as those of Bermuda grass and species of bindweed) may persist under the barrier and within the soil seed bank. Termites are attracted to the area. While they are a natural part of the ecosystem that transforms the weed barrier into rich soil, they can pose a hazard to nearby wood-framed structures.

- Slug populations may increase during the early stages of decomposition. However they can be kept away or harvested. The system may need a constant supply of organic material, at least during the early stages. Roaming animals may interrupt the sheet mulching process.

Vegan Organic Gardening

Vegan organic gardening and farming is the organic cultivation and production of food crops and other crops with a minimal amount of exploitation or harm to any animal. Vegan gardening and stock-free farming methods use no animal products or by-products, such as bloodmeal, fish products, bone meal, feces, or other animal-origin matter, because the production of these materials is viewed as either harming animals directly, or being associated with the exploitation and consequent suffering of animals. Some of these materials are by-products of animal husbandry, created during the process of cultivating animals for the production of meat, milk, skins, furs, entertainment, labor, or companionship; the sale of by-products decreases expenses and increases profit for those engaged in animal husbandry, and therefore helps support the animal husbandry industry, an outcome most vegans find unacceptable.

Types

Forest Gardening

Forest gardening is a fully plant-based organic food production system based on woodland ecosystems, incorporating fruit and nut trees, shrubs, herbs, vines and perennial vegetables. Making use of companion planting, these can be intermixed to grow in a succession of layers, to replicate a woodland habitat. Forest gardening can be viewed as a way to recreate the Garden of Eden. The three main products from a forest garden are fruit, nuts and green leafy vegetables.

Robert Hart adapted forest gardening for temperate zones during the early 1960s. Robert Hart began with a conventional smallholding at Wenlock Edge in Shropshire. However, following his adoption of a raw vegan diet for health and personal reasons, Hart replaced his farm animals with plants. He created a model forest garden from a small orchard on his farm and intended naming his gardening method or. Hart later dropped these terms once he became aware that and were already being used to describe similar systems in other parts of the world.

Robert Hart's forest garden in Shropshire, England.

Vegan Permaculture

Vegan permaculture (also known as veganic permaculture, veganiculture, or vegaculture) avoids the use of domesticated animals. It is essentially the same as permaculture except for the addition of a fourth core value; "Animal Care". Zalan Glen, a raw vegan, proposes that should emerge from permaculture in the same way veganism split from vegetarianism in the 1940s. Vegan permaculture recognizes the importance of free-living animals, not domesticated animals, to create a balanced ecosystem.

Veganic Gardening

The veganic gardening method is a distinct system developed by Rosa Dalziell O'Brien, Kenneth Dalziel O'Brien and May E. Bruce, although the term was originally coined by Geoffrey Rudd as a contraction of in order to "denote a clear distinction between conventional chemical based systems and organic ones based on animal manures". The O'Brien system's principal argument is that animal manures are harmful to soil health rather than that their use involves exploitation of and cruelty to animals.

The system employs very specific techniques including the addition of straw and other vegetable wastes to the soil in order to maintain soil fertility. Gardeners following the system use soil-covering mulches, and employ non-compacting surface cultivation techniques using any short-handled, wide-bladed, hand hoe. They kneel when surface cultivating, placing a board under their knees to spread out the pressure, and prevent soil compaction. Kenneth Dalziel O'Brien published a description of his system:

> "The veganic method of clearing heavily infested land is to take advantage of a plant's tendencies to move its roots nearer to the soil's surface when it is deprived of light. To make use of this principle, aided by a decaying process of the top growth of weeds, etc., it is necessary to subject such growth to heat and moisture in order to speed up the decay, and this

is done by applying lime, then a heavy straw cover, and then the herbal compost activator The following are required: Sufficient new straw to cover an area to be cleared to a depth of 3 to 4 inches".

The O'Brien method also advocates minimal disturbance of the soil by tilling, the use of cover crops and green manures, the creation of permanent raised beds and permanent hard-packed paths between them, the alignment of beds along a north-south axis, and planting in double rows or more so that not every row has a path on both sides. Use of animal manure is prohibited.

Vegan Biodynamic Agriculture

The German agricultural researcher Maria Thun (1922 - 2012) developed vegan equivalents to the traditional, animal based biodynamic preparations. As a reaction to the BSE scandal in Europe, she started researching plant based preparations, using tree barks as replacement for animal organs as sheath for the preparations.

In particular in Italy, there is a movement of vegan biodynamic farming, represented by farmers such as Sebastiano Cossia Castiglioni and Cristina Menicocci.

There are many other methods currently used and under development. However, to be certified DEMETER BIODYNAMIC the regular BD preparations must be used. Because the BD preparations require the slaughtering of deer and cows and BD preps must be used in the compost for soil amendments, sprayed on the fields, the DEMETER certified products cannot claim to be vegan or vegetarian.

Practices

Soil fertility is maintained by the use of green manures, cover crops, green wastes, composted vegetable matter, and minerals. Some vegan gardeners may supplement this with human urine from vegans (which provides nitrogen) and 'humanure' from vegans, produced from compost toilets. Generally only waste from vegans is used because of the expert recommendation that the risks associated with using composted waste are acceptable only if the waste is from animals or humans having a largely herbivorous diet.

Veganic gardeners may prepare soil for cultivation using the same method used by conventional and organic gardeners of breaking up the soil with hand tools and power tools and allowing the weeds to decompose.

Companion planting in gardening and agriculture is the planting of different crops in proximity for any of a number of different reasons, including pest control, pollination, providing habitat for beneficial insects, maximizing use of space, and to otherwise increase crop productivity. Companion planting is a form of polyculture.

Companion planting is used by farmers and gardeners in both industrialized and developing countries for many reasons. Many of the modern principles of companion planting were present many centuries ago in cottage gardens in England and forest gardens in Asia, and thousands of years ago in Mesoamerica.

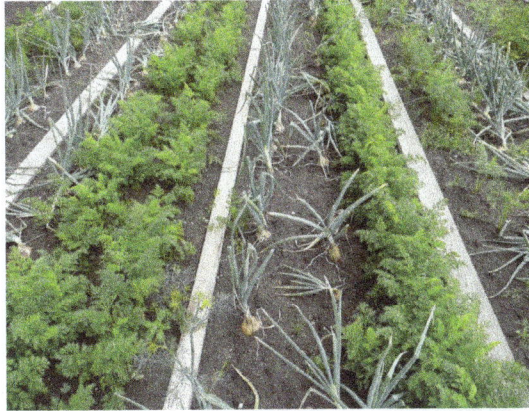
Companion planting of carrots and onions.

Companion Planting

Companion planting in gardening and agriculture is the planting of different crops in proximity for any of a number of different reasons, including pest control, pollination, providing habitat for beneficial insects, maximizing use of space, and to otherwise increase crop productivity. Companion planting is a form of polyculture.

Companion planting is used by farmers and gardeners in both industrialized and developing countries for many reasons. Many of the modern principles of companion planting were present many centuries ago in cottage gardens in England and forest gardens in Asia, and thousands of years ago in Mesoamerica.

Mechanisms

Companion planting can operate through a variety of mechanisms, which may sometimes be combined.

Provision of Nutrients

Legumes such as clover provide nitrogen compounds to other plants such as grasses by fixing nitrogen from the air with symbiotic bacteria in their root nodules.

Dandelions have long taproots that bring nutrients from deep within the soil to near the surface, benefitting neighboring plants that are shallower-rooted.

Trap Cropping

Trap cropping uses alternative plants to attract pests away from a main crop. For example, nasturtium is a food plant of some caterpillars which feed primarily on members of the cabbage family (brassicas); some gardeners claim that planting them around brassicas protects the food crops from damage, as eggs of the pests are preferentially laid on the nasturtium. However, while many trap crops have successfully diverted pests off of focal crops in small scale greenhouse, garden and field experiments, only a small portion of these plants have been shown to reduce pest damage at larger commercial scales.

Host-finding Disruption

Recent studies on host-plant finding have shown that flying pests are far less successful if their host-plants are surrounded by any other plant or even "decoy-plants" made of green plastic, cardboard, or any other green material.

The host-plant finding process occurs in phases:

- The first phase is stimulation by odours characteristic to the host-plant. This induces the insect to try to land on the plant it seeks. But insects avoid landing on brown (bare) soil. So if only the host-plant is present, the insects will quasi-systematically find it by simply landing on the only green thing around. This is called (from the point of view of the insect) "appropriate landing". When it does an "inappropriate landing", it flies off to any other nearby patch of green. It eventually leaves the area if there are too many 'inappropriate' landings. The second phase of host-plant finding is for the insect to make short flights from leaf to leaf to assess the plant's overall suitability. The number of leaf-to-leaf flights varies according to the insect species and to the host-plant stimulus received from each leaf. The insect must accumulate sufficient stimuli from the host-plant to lay eggs; so it must make a certain number of consecutive 'appropriate' landings. Hence if it makes an 'inappropriate landing', the assessment of that plant is negative, and the insect must start the process anew. Thus it was shown that clover used as a ground cover had the same disruptive effect on eight pest species from four different insect orders. An experiment showed that 36% of cabbage root flies laid eggs beside cabbages growing in bare soil (which resulted in no crop), compared to only 7% beside cabbages growing in clover (which allowed a good crop). Simple decoys made of green cardboard also disrupted appropriate landings just as well as did the live ground cover.

Pest Suppression

Some companion plants help prevent pest insects or pathogenic fungi from damaging the crop, through chemical means. For example, the smell of the foliage of marigolds is claimed to deter aphids from feeding on neighbouring plants.

Predator Recruitment

Companion plants that produce copious nectar or pollen in a vegetable garden (insectary plants) may help encourage higher populations of beneficial insects that control pests, as some beneficial predatory insects only consume pests in their larval form and are nectar or pollen feeders in their adult form. For instance, marigolds with simple flowers attract nectar-feeding adult hoverflies, the larvae of which are predators of aphids.

Protective Shelter

Some crops are grown under the protective shelter of different kinds of plant, whether as wind breaks or for shade. For example, shade-grown coffee, especially, has traditionally been grown in light shade created by scattered trees with a thin canopy, allowing light through to the coffee bushes but protecting them from overheating. Suitable Asian trees include (tton tong or dadap), (khae falang), (khi lek), (khao dao sang), and, a useful timber tree.

Shade-grown coffee plantation in Costa Rica. The red trees in the background provide shade; those in the foreground have been pruned to allow full exposure to the sun.

Systems

Systems in use or being trialled include:

- Square foot gardening attempts to protect plants from many normal gardening problems, such as weed infestation, by packing them as closely together as possible, which is facilitated by using companion plants, which can be closer together than normal.

- Forest gardening, where companion plants are intermingled to create an actual ecosystem, emulates the interaction of up to seven levels of plants in a forest or woodland.

- Organic gardening makes frequent use of companion planting, since many other means of fertilizing, weed reduction and pest control are forbidden.

Waru Waru

Waru Waru is and agricultural technique that was developed in South America. It consists of raised beds and irrigation to prevent soil erosion from doing damage during floods. This technique helps to collect water but at the same time drain it so that it will not be effected by brutal floods.

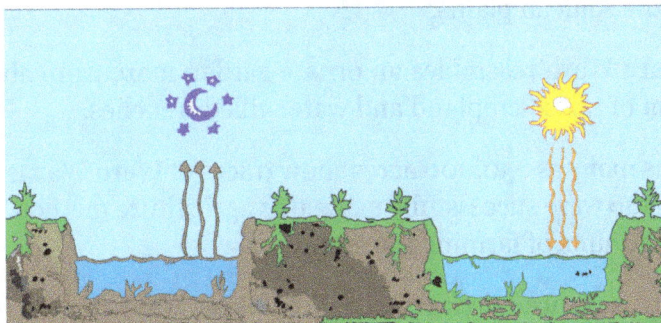

Thousands of years ago, our ancestors came up with efficient farming techniques that modern scientists are now investigating to mitigate climate change. One of the most interesting technologies of ancient time are underground irrigation canals known as "qanats", developed by people in Persia about 3,000 years ago. The qanats transported water from a water wells to surface for irrigation and drinking as well and they could do it over long distances.

The Hohokam people were 'Masters of the Desert'. Their extensive irrigation canals served as the foundation of the modern canals used to irrigate crops in the valleys along the Salt and Gila rivers. The Aztecs built so-called chinampas to improve their agriculture. These were small, artificial islands created on a freshwater lake. The chinampas resembled floating gardens.

Chinampas: Artificial Islands created by the aztecs to improve agriculture.

Chinampas were used throughout the Valley of Mexico around the lake bed and were without doubt one of the reasons why Aztec's farming became famous.

Knowledge of Waru Waru is not Lost

Before the rise of the Inca Empire, Andean people developed an agricultural technique called Waru Waru. This technology, based on modification of the soil surface to facilitate water movement and storage helped people to cope with floods and droughts.

Combining raised beds with irrigation channels to prevent damage by soil erosion during floods, was an inexpensive way to improve crop yields and ease the punishing effects of farming at 12,500 feet above sea level on the Andean plains.

Also known as , the Waru Waru resembles an ornate garden maze from above. It's a cleverly designed patterned system of raised cropland and water-filled trenches.

Around Lake Titicaca, as many as 250,000 acres show traces of Waru Warus. This suggests ancient inhabitants of the altiplano were successfully using this agriculture method to tackle the considerable environmental constraints of farming the area.

When more technically advanced irrigation technologies were discovered, Waru Waru was abandoned. However, in more recent years, the Waru Waru technology has been brought back to life in Bolivia and Peru. Scientists discovered that Waru Waru technology not only worked, but it tripled crop production.

In 1981, Clark Erickson, of the University of Illinois, recognized the archaeological significance of Waru Waru. Erickson wondered whether this ancient system might not serve modern farmers and began to rebuild some of the raised fields. Using traditional Andean tools, local farmers planted an experimental field with potatoes, quinoa and canihua. Private development organizations and the Peruvian government rushed in to aid farmers.

Locals in Cutini Capilla, an Aymara community on the western shore of Lake Titicaca between Puno and the ruins of the Tiwanaku capital said the Waru Warus have helped them to make the land useful again.

Waru Waru technology.

Alipio Canahua, an agronomist working with the Food and Agriculture Organisation (FAO), says that the ancient agricultural system, which could date back 3,000 years, actually creates its own microclimate.

"It captures water when there are droughts and drains away water when there's too much rain, meaning that it irrigates the crops all year round". "When it comes to temperature, we've measured a rise of three degrees centigrade in the immediate environment around it – this can save a significant percentage of the crops from being killed in frosts".

The ancient Waru Waru technology has been brought back to life, but whether it survives depends on investments and financial support from the governments.

It is also impossible to make the suqakollos as big as pre-Hispanic people did because roads and boundaries between communal lands, have limited the space available.

Scientists think that the ancient Waru Waru technology could be used in conjunction with other ancient techniques such as interconnected irrigation lakes called "qochas" and the Inca-built terraces known as "andenes" for farming the steep Andean slopes.

Raised Beds and Waru Waru Cultivation

This technology is based on modification of the soil surface to facilitate water movement and storage, and to increase the organic content of the soil to increase its suitability for cultivation. This system of soil management for irrigation purposes was first developed in the year 300 B.C., before the rise of the Inca Empire.

The technology is a combination of rehabilitation of marginal soils, drainage improvement, water storage, optimal utilization of available radiant energy, and attenuation of the effects of frost. The main feature of this system is the construction of a network of embankments and canals. The embankments serve as raised beds for cultivation of crops, while the canals are used for water storage and to irrigate the plants. The soils used for the embankments are compacted to facilitate water retention by reducing porosity, permeability, and infiltration. Infiltration in the clay soils of the region varies from 20% to 30% of the precipitation volume. Thus, clay soils are preferred for this purpose. Sandy soils have too great a porosity to retain the water within the beds.

The cultivation takes place in the "new" soils within the raised bed created by the construction of the embankment. Within the bed, the increased porosity of the new soils results in enhanced infiltration, often increasing infiltration by 80% to 100% of the original soil. This system permits the recycling of nutrients and and all the other chemical and biological processes necessary for crop production. Water uptake by the raised beds is through diffusion and capillary movements using water contained within the beds or supplied from the surrounding canals. The soils are kept at an adequate moisture level to facilitate the cultivation of plants such as potatoes and quinoa Thermal energy is captured and retained in the soil as a result of the enhanced moisture levels, which protect the soils of the bed from the effects of frost. The system acts as a thermoregulator of the microclimate within the bed.

There are three types of raised bed systems, characterized by the source of water:

- Rainwater systems, in which rainwater is the primary source of moisture. These systems require small lagoons for storage during dry periods and a system of canals to distribute the water to the beds. They are usually located at the base of a hill or a mountain.

- Fluvial systems, in which moisture is supplied by water from nearby rivers. These systems require a hydraulic infrastructure, such as canals and dikes, to transport the water.

- Phreatic systems, in which groundwater is the source of moisture in the beds. These systems are located in areas where the groundwater table is close to the surface of the soil and there is a mechanism for groundwater recharge, such as an infiltration lagoon.

The main design considerations for raised bed cultivation include the following:

- Depth of the water table, since a high water table increases the height of the embankment required.

- Soil characteristics, which affect both the dimensions of the embankment and the nature of the cultivation zone.

- Climatic conditions, include the volume and frequency of rainfall, temperature range, and frost frequency.

An example of a typical embankment and canal system is shown in figure. Soft fill (e.g., compost or mulch) might be required within the embanked bed to maintain an adequate level of soil moisture.

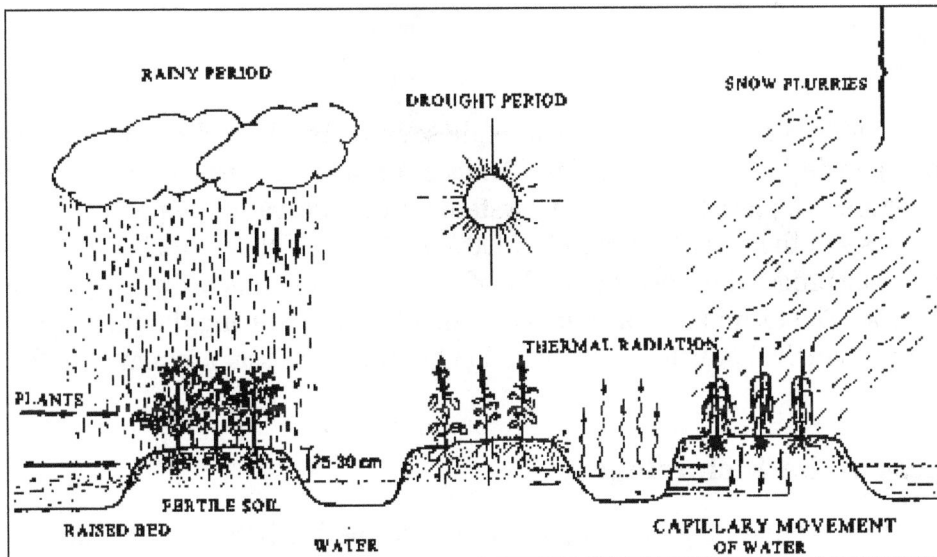

Raised Bed Irrigation System in Puno, Peru.

Extent of Use

This technology has been used primarily in the Lake Titicaca region at Puno, Peru, and in the Illpa River basin of Bolivia.

Operation and Maintenance

Periodic reconstruction of the embankments or raised beds is necessary to repair damage caused by erosion and water piping. Reconstruction is usually done during the dry season (March to May, in Peru), although in some areas it is done immediately after harvesting because of a lack of available labor at other times of the year. Cultivation of pasture and other grasses of differing heights on the embankments will help to prevent or control erosion caused by torrential rains during the wet season. Cultivation practices can also damage the embankments. Raising animals such as hogs near the embankments should be avoided, since they can damage the cultivation areas in their search for food.

Periodic fertilization of the raised beds is recommended, and the use of insecticides and fungicides may be necessary to limit crop damage. Insecticides are particularly advisable in the cultivation of potatoes.

Level of Involvement

This technology has been promoted, and assistance to farmers provided, by several Peruvian governmental organizations, including the Institute Nacional de Investigación Agropecuaria y Agroindustrial (INIAA), the Centre de Investigación Agropecuaria Salcedo (CIAS), the Centro de Proyectos Integrales Andinos (CEPIA), and by a number of NGOs. These organizations intend to reconstruct 500 ha of in 72 rural communities in the vicinity of Puno. Such an approach is considered to be representative of the involvement necessary to successfully implement a cultivation program in the region. Once established, the operation and maintenance of the systems, like the planting and harvesting of agricultural products, becomes the responsibility of the farmers who benefit from the use of this technology.

Costs

Very little information is available on the costs of these systems. The technology is at present largely experimental and limited to portions of the Andean Altiplano in Peru and Bolivia. Nevertheless, the cost per hectare of a phreatic raised-bed system for the cultivation of potatoes is estimated at $1 460 on the basis of the system created in Chatuma, Peru. Of this, 70% is direct cost and 30% is indirect cost. The production cost for 11.2 kg of potatoes using this technology in Chatuma was estimated at $480. The technology produces economic benefits during the first 3 years following construction, but, shortly thereafter reconstruction becomes necessary to maintain the productivity of the system.

Effectiveness of the Technology

In the communities around Puno, during the seven-year period between 1982 and 1989, 229 ha were converted to this technology, with mixed results. Some areas experienced large increases in productivity, particularly in the cultivation of potatoes, while other areas did not. Climatic conditions, such as drought and extremely cold weather, are likely to have contributed to the decrease in productivity in some areas, while poor design and construction of embankments may have led to the decline in productivity recorded in others.

Suitability

This technology is suitable in areas with extreme climatic conditions, such as mountainous areas that experience heavy rainfalls and periodic droughts, and where temperature fluctuations range from intense heat to frost. It should be very useful in arid and semi-arid areas.

Advantages

- This technology can contribute to mitigating the effects of extreme climatic variations.
- The construction cost is relatively low.
- It can increase the production of certain agricultural crops.

Disadvantages

- The life span of the technology is relatively short; the systems require reconstruction after about 3 years of operation.
- Testing of soil texture and composition is necessary before implementation.
- systems require annual maintenance and periodic repair.

Cultural Acceptability

This is an ancient technology, well accepted in the agricultural communities of Peru and Bolivia.

Further Development of the Technology

Application of this technology in other areas with different soil and climatic conditions will be a measure of its potential utility outside of the areas where it is traditionally used. Improvements in the design of the raised bed cultivation system are necessary in order to extend the economic life of the technology and to minimize the need for regular reconstruction of the beds to maintain their productivity.

References

- Jacke, Dave, and Toensmeier, Eric 2005. Edible Forest Gardens. Two volume set. Volume One: Ecological Vision and Theory for Temperate Climate Permaculture, ISBN 1-931498-79-2. Volume Two: Ecological Design and Practice for Temperate Climate Permaculture, ISBN 1-931498-80-6. White River Junction, VT: Chelsea Green
- Stamets, Paul (2005). Mycelium running: how mushrooms can help save the world. Ten Speed Press. ISBN 1-58008-579-2
- Waru-waru-ancient-andean-irrigation-system-brought-back-to-life: ancientpages.com, Retrieved 14 February, 2019
- What-is-intercropping: worldatlas.com, Retrieved 28 June, 2019
- 4-types-of-intercropping-in-agriculture: 1001artificialplants.com, Retrieved 25 July, 2019
- 6-advantages-and-disadvantages-of-intercropping: connectusfund.org, Retrieved 24 April, 2019

Other Fields Related to Permaculture

- **Organic Farming**

- **Integrated Farming**

- **Community Gardening**

- **Sustainable Development**

- **Rainwater Harvesting**

Permaculture is a vast subject that finds its applications in various other fields. Some of them include organic farming, integrated farming, community gardening, sustainable development, rainwater harvesting, etc. This chapter has been carefully written to provide an easy understanding of the varied fields related to permaculture.

Organic Farming

Organic farming is a technique, which involves cultivation of plants and rearing of animals in natural ways. This process involves the use of biological materials, avoiding synthetic substances to maintain soil fertility and ecological balance thereby minimizing pollution and wastage. In other words, organic farming is a farming method that involves growing and nurturing crops without the use of synthetic based fertilizers and pesticides. Also, no genetically modified organisms are permitted.

It relies on ecologically balanced agricultural principles like crop rotation, green manure, organic waste, biological pest control, mineral and rock additives. Organic farming make use of pesticides and fertilizers if they are considered natural and avoids the use of various petrochemical fertilizers and pesticides.

Differences between Organic and Conventional Farming Methods

In conventional farming method, before seeds are sown, the farmer will have to treat or fumigate his farm using harsh chemicals to exterminate any naturally existing fungicides. He will fertilize the soil using petroleum based fertilizers. On the flip side, the organic farmer will prepare and enrich his land before sowing by sprinkling natural based fertilizers such as manure, bone meal or shellfish fertilizer.

Before planting seeds, the organic farmer will soak the seeds in fungicides and pesticides to keep insects and pests at bay. Chemical are also incorporated in the irrigation water to prevent insects from stealing the planted seeds. On the other hand, the organic farmer will not soak his seeds in any chemical solution nor irrigate the newly planted seeds using water with added chemicals. In fact, he will not even irrigate with council water, which is normally chlorinated to kill any bacteria. He will depend on natural rain or harvest and stored rainwater to use during dry months.

When the seeds have sprung up, and it's time to get rid of weeds, the conventional farmer will use weedicide to exterminate weeds. The organic farmer will not use such chemicals to get rid of the weed problem. Instead, he will physically weed out the farm, although it's very labor intensive. Better still, the organic farmer can use a flame weeder to exterminate weeds or use animals to eat away the weeds.

When it comes to consumption, it's a no-brainer that anyone consuming products from the conventional farmer will absorb the pesticide and weedicide residues into the body, which could lead to developing dangerous diseases like cancer. People understand that health is important to them and that's why they are going organic in record numbers today.

Reasons for Organic Farming

The population of the planet is skyrocketing and providing food for the world is becoming extremely difficult. The need of the hour is sustainable cultivation and production of food for all. The Green Revolution and its chemical based technology are losing its appeal as dividends are falling and returns are unsustainable. Pollution and climate change are other negative externalities caused by use of fossil fuel based chemicals.

In spite of our diet choices, organic food is the best choice you'll ever make, and this means embracing organic farming methods. Here are the reasons why we need to take up organic farming methods:

1. To accrue the benefits of nutrients: Foods from organic farms are loaded with nutrients such as vitamins, enzymes, minerals and other micro-nutrients compared to those from conventional farms. This is because organic farms are managed and nourished using sustainable practices. In fact, some past researchers collected and tested vegetables, fruits, and grains from both organic farms and conventional farms.

The conclusion was that food items from organic farms had way more nutrients than those sourced from commercial or conventional farms. The study went further to substantiate that five servings of these fruits and vegetables from organic farms offered sufficient allowance of vitamin C. However, the same quantity of fruits and vegetable did not offer the same sufficient allowance.

2. Stay away from GMOs: Statistics show that genetically modified foods (GMOs) are contaminating natural foods sources at real scary pace, manifesting grave effects beyond our comprehension. What makes them a great threat is they are not even labeled. So, sticking to organic foods sourced from veritable sources is the only way to mitigate these grave effects of GMOs.

3. Natural and better taste: Those that have tasted organically farmed foods would attest to the fact that they have a natural and better taste. The natural and superior taste stems from the well balanced and nourished soil. Organic farmers always prioritize quality over quantity.

4. Direct support to farming: Purchasing foods items from organic farmers is a surefire investment in a cost-effective future. Conventional farming methods have enjoyed great subsidies and tax cuts from most governments over the past years. This has led to the proliferation of commercially produced foods that have increased dangerous diseases like cancer. It's time governments invested in organic farming technologies to mitigates these problems and secure the future. It all starts with you buying food items from known organic sources.

5. To conserve agricultural diversity: These days, it normal to hear news about extinct species and this should be a major concern. In the last century alone, it is approximated that 75 percent of agricultural diversity of crops has been wiped out. Slanting towards one form of farming is a recipe for disaster in the future. A classic example is a potato. There were different varieties available in the marketplace. Today, only one species of potato dominate.

This is a dangerous situation because if pests knock out the remaining potato specie available today, we will not have potatoes anymore. This is why we need organic farming methods that produce disease and pest resistant crops to guarantee a sustainable future.

6. To prevent antibiotics, drugs, and hormones in animal products: Commercial dairy and meat are highly susceptible to contamination by dangerous substances. A statistic in an American journal revealed that over 90% of chemicals the population consumes emanate from meat tissue and dairy products. According to a report by Environmental Protection Agency (EPA), a vast majority of pesticides are consumed by the population stem from poultry, meat, eggs, fish and dairy product since animals and birds that produce these products sit on top of the food chain.

This means they are fed foods loaded with chemicals and toxins. Drugs, antibiotics, and growth hormones are also injected into these animals and so, are directly transferred to meat and dairy products. Hormone supplementation fed to farmed fish, beef and dairy products contribute mightily to ingestion of chemicals. These chemicals only come with a lot of complications like genetic problems, cancer risks, growth of tumor and other complications at the outset of puberty.

Key Features of Organic Farming

- Protecting soil quality using organic material and encouraging biological activity.
- Indirect provision of crop nutrients using soil microorganisms.

- Nitrogen fixation in soils using legumes.

- Weed and pest control based on methods like crop rotation, biological diversity, natural predators, organic manures and suitable chemical, thermal and biological intervention.

- Rearing of livestock, taking care of housing, nutrition, health, rearing and breeding.

- Care for the larger environment and conservation of natural habitats and wildlife.

Four Principles of Organic Farming

- Principle of Health: Organic agriculture must contribute to the health and well being of soil, plants, animals, humans and the earth. It is the sustenance of mental, physical, ecological and social well being. For instance, it provides pollution and chemical free, nutritious food items for humans.

- Principle of Fairness: Fairness is evident in maintaining equity and justice of the shared planet both among humans and other living beings. Organic farming provides good quality of life and helps in reducing poverty. Natural resources must be judiciously used and preserved for future generations.

- Principle of Ecological Balance: Organic farming must be modeled on living ecological systems. Organic farming methods must fit the ecological balances and cycles in nature.

- Principle of Care: Organic agriculture should be practiced in a careful and responsible manner to benefit the present and future generations and the environment.

As opposed to modern and conventional agricultural methods, organic farming does not depend on synthetic chemicals. It utilizes natural, biological methods to build up soil fertility such as microbial activity boosting plant nutrition.

Secondly, multiple cropping practiced in organic farming boosts biodiversity which enhances productivity and resilience and contributes to a healthy farming system. Conventional farming systems use mono cropping that destroys the soil fertility.

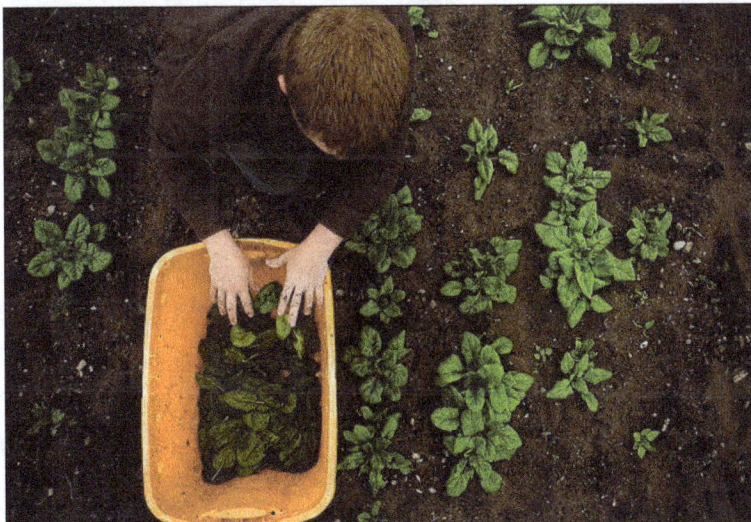

Modern Farming is Unsustainable because

- Loss of soil fertility due to excessive use of chemical fertilizers and lack of crop rotation.

- Nitrate run off during rains contaminates water resources.

- Soil erosion due to deep ploughing and heavy rains.

- More requirement of fuel for cultivation.

- Use of poisonous bio-cide sprays to curb pest and weeds.

- Cruelty to animals in their housing, feeding, breeding and slaughtering.

- Loss of biodiversity due to mono culture.

- Native animals and plants lose space to exotic species and hybrids.

Benefits of Organically Grown Food Items and Agricultural Produce

- Better Nutrition: As compared to a longer time conventionally grown food, organic food is much richer in nutrients. Nutritional value of a food item is determined by its mineral and vitamin content. Organic farming enhances the nutrients of the soil which is passed on to the plants and animals.

- Helps us stay healthy: Organic foods do not contain any chemical. This is because organic farmers don't use chemicals at any stage of the food-growing process like their commercial counterparts. Organic farmers use natural farming techniques that don't harm humans and environment. These foods keep dangerous diseases like cancer and diabetes at bay.

- Free of poison: Organic farming does not make use of poisonous chemicals, pesticides and weedicides. Studies reveal that a large section of the population fed on toxic substances used in conventional agriculture have fallen prey to diseases like cancer. As organic farming avoids these toxins, it reduces the sickness and diseases due to them.

- Organic foods are highly authenticated: For any produce to qualify as organic food, it must undergo quality checks and the creation process rigorously investigated. The same rule applies to international markets. This is a great victory for consumers because they are getting the real organic foods. These quality checks and investigations weed out quacks who want to benefit from the organic food label by delivering commercially produced foods instead.

- Lower prices: There is a big misconception that organic foods are relatively expensive. The truth is they are actually cheaper because they don't require application of expensive pesticides, insecticides, and weedicides. In fact, you can get organic foods direct from the source at really reasonable prices.

- Enhanced Taste: The quality of food is also determined by its taste. Organic food often tastes better than other food. The sugar content in organically grown fruits and vegetables provides them with extra taste. The quality of fruits and vegetables can be measured using Brix analysis.

- Organic farming methods are eco-friendly: In commercial farms, the chemicals applied infiltrate into the soil and severely contaminate it and nearby water sources. Plant life, animals, and humans are all impacted by this phenomenon. Organic farming does not utilize these harsh chemicals so; the environment remains protected.

- Longer shelf–life: Organic plants have greater metabolic and structural integrity in their cellular structure than conventional crops. This enables storage of organic food for a longer time.

Organic farming is preferred as it battles pests and weeds in a non-toxic manner, involves less input costs for cultivation and preserves the ecological balance while promoting biological diversity and protection of the environment.

Integrated Farming

Integrated Farming or Integrated Farm Management is a whole farm management system which aims to deliver more sustainable agriculture. It is a dynamic approach which can be applied to any farming system around the world. It involves attention to detail and continuous improvement in all areas of a farming business through informed management processes. Integrated Farming combines the best of modern tools and technologies with traditional practices according to a given site and situation. In simple words, it means using many ways of cultivation in a small space or land.

The holistic approach UNI 11233 new European bio standard: Integrated production
system looks at and relates to the whole Organic and Bio farm.

The International Organisation of Biological Control (IOBC) describes Integrated Farming as a farming system where high quality organic food, feed, fibre and renewable energy are produced by using resources such as soil, water, air and nature as well as regulating factors to farm sustainably and with as little polluting inputs as possible.

Particular emphasis is placed on an integrated organic management approach looking at the whole Bio farm as cross-linked unit, on the fundamental role and function of agro-ecosystems, on nutrient cycles which are balanced and adapted to the demand of the crops, and on health and welfare of all livestock on the farm. Preserving and enhancing soil fertility, maintaining and improving a diverse environment and the adherence to ethical and social criteria are indispensable basic elements. Crop protection takes into account all biological, technical and chemical methods which then are balanced carefully and with the objective to protect the environment, to maintain profitability of the business and fulfil social requirements.

EISA European Initiative for Sustainable Development in Agriculture e. V. have an Integrated Farming Framework which provides additional explanations on key aspects of Integrated Farming. These include: Organisation & Planning, Human & Social Capital, Energy Efficiency, Water Use & Protection, Climate Change & Air Quality, Soil Management, Crop Nutrition, Crop Health & Protection, Animal Husbandry, Health & Welfare, Landscape & Nature Conservation and Waste Management Pollution Control.

LEAF (Linking Environment and Farming) in the UK promotes a comparable model and defines Integrated Farm Management (IFM) as whole farm business approach that delivers more sustainable farming. LEAF's Integrated Farm Management consists of nine interrelated sections: Organisation & Planning, Soil Management & Fertility, Crop Health & Protection, Pollution Control & By-Product Management, Animal Husbandry, Energy Efficiency, Water Management, and Landscape & Nature Conservation.

Classification

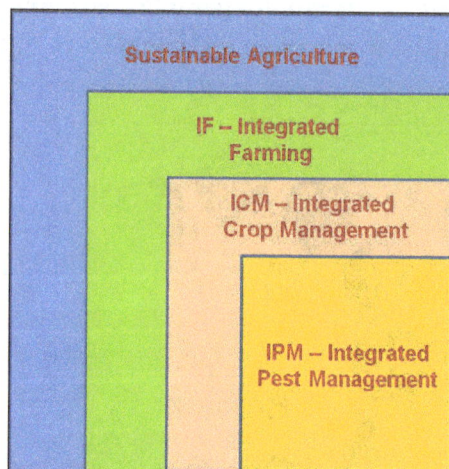

Integrated farming in the context of sustainable agriculture.

The Food and Agriculture Organization of the United Nations FAO promotes Integrated Pest Management (IPM) as the preferred approach to crop protection and regards it as a pillar of both sustainable intensification of crop production and pesticide risk reduction. IPM thus is one indispensable element of Integrated Crop Management which in turn is one essential part of the holistic Integrated Farming approach towards sustainable agriculture.

KELLER, 1986 highlights that Integrated Crop Management is not to be understood as compromise between different agricultural production systems. It rather must be understood as production system with a targeted, dynamic and continuous use and development of experiences which were made in the so-called conventional farming. In addition to natural scientific findings, impulses from organic farming are also taken up.

Objectives

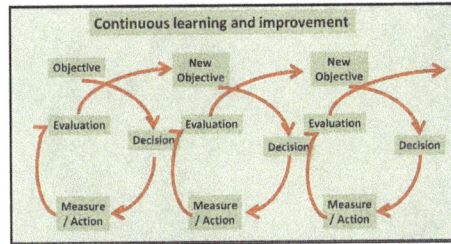

Continuous learning process in Integrated Farming.

Integrated Farming is based on attention to detail, continuous improvement and managing all resources available.

Being bound to sustainable development, the underlying three dimensions economic development, social development and environmental protection are thoroughly considered in the practical implementation of Integrated Farming. However, the need for profitability is a decisive prerequisite: To be sustainable, the system must be profitable, as profits generate the possibility to support all activities outlined in the (EISA Integrated Farming) IF Framework.

As a management and planning approach, Integrated Farming includes regular benchmarking of targets set against results achieved. The concept of the EISA Integrated Farming Framework for example has a clear focus on farmers' awareness of their own performance. By regularly benchmarking their performance, farmers become aware of achievements as well as deficiencies, and by paying attention to detail they can continuously work on improving the whole farming enterprise and their economic performance at the same time: According to findings in UK, reducing fertiliser and chemical inputs to amounts according to the demand of the crops allowed for cost savings in the range of £2,500 – £10,000 per year and per farm.

Community Gardening

A community garden is a single piece of land gardened collectively by a group of people. Community gardens utilize either individual or shared plots on private or public land while producing fruit, vegetables, and plants grown for their attractive appearance. Around the world the community gardens can fulfill variety of purposes such as aesthetic and community improvement, physical or mental well-being, or land conservation.

Strathcona Heights Community Garden in Ottawa, Canada.

Purpose

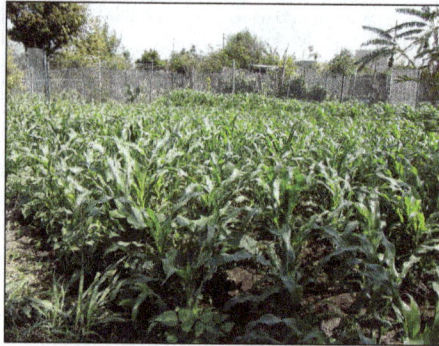

Crops at the former South Central Farm in Los Angeles, California, United States.

According to Marin Master Gardeners, "a community garden is any piece of land gardened by a group of people, utilizing either individual or shared plots on private or public land". Community gardens provide fresh products and plants as well as contributing to a sense of community and connection to the environment and an opportunity for satisfying labor and neighborhood improvement. They are publicly functioning in terms of ownership, access, and management, as well as typically owned in trust by local governments or not for profit associations.

Community gardens vary widely throughout the world. In North America, community gardens range from familiar "victory garden" areas where people grow small plots of vegetables, to large "greening" projects to preserve natural areas, to tiny street beautification planters on urban street corners. Some grow only flowers, others are nurtured communally and their bounty shared. There are even non-profits in many major cities that offer assistance to low-income families, children groups, and community organizations by helping them develop and grow their own gardens. In the UK and the rest of Europe, closely related "allotment gardens" can have dozens of plots, each measuring hundreds of square meters and rented by the same family for generations. In the developing world, commonly held land for small gardens is a familiar part of the landscape, even in urban areas, where they may function as market gardens. They also practice crop rotations with versatile plants such as peanuts, tomatoes and much more.

Community gardens are often used in urban neighborhoods to alleviate the food desert effect. Food accessibility described in urban areas refers to residents who have limited access to fresh produce such as fruits and vegetables. Food deserts often serve lower-income neighborhoods usually in which residents are forced to rely on unhealthy food options such as expensive processed foods from convenience stores, gas stations, and fast-food restaurants. Community gardens provide accessibility for fresh food to be in closer proximity located in local neighborhoods. Community gardens can help expand the realm for ensuring residents' access to healthy and affordable food in a community.

Community gardens may help alleviate one effect of climate change, which is expected to cause a global decline in agricultural output, making fresh produce increasingly unaffordable. Community gardens are also an increasingly popular method of changing the built environment in order to promote health and wellness in the face of urbanization. The built environment has a wide range of positive and negative effects on the people who work, live, and play in a given area, including a person's chance of developing obesity Community gardens encourage an urban community's food security, allowing citizens to grow their own food or for others to donate what they have grown. Advocates say locally grown food decreases a community's reliance on fossil fuels for transport of food from large agricultural areas and reduces a society's overall use of fossil fuels to drive in agricultural machinery. A 2012 op-ed by community garden advocate Les Kishler examines how community gardening can reinforce the so-called "positive" ideas and activities of the Occupy movement.

Community gardens improve users' health through increased fresh vegetable consumption and providing a venue for exercise. A fundamental part of good health is a diet rich in fresh fruits, vegetables, and other plant based foods. Community gardens provide access to such foods for the communities in which they are located. Community gardens are especially important in communities with large concentrations of low socioeconomic populations, as a lack fresh fruit and vegetable availability plagues these communities at disproportionate rates.

The gardens also combat two forms of alienation that plague modern urban life, by bringing urban gardeners closer in touch with the source of their food, and by breaking down isolation by creating a social community. Community gardens provide other social benefits, such as the sharing of food production knowledge with the wider community and safer living spaces. Active communities experience less crime and vandalism.

Types of Gardens

A 20ft x 20ft community garden plot in Harrisonburg, Virginia.

There are multiple types of community gardens with distinct varieties in which the community can participate in.

- Neighborhood gardens are the most common type that is normally defined as a garden where a group of people come together to grow fruits, vegetables and ornamentals. They are identifiable as a parcel of private or public land where individual plots are rented by gardeners at a nominal annual fee.

- Residential Gardens are typically shared among residents in apartment communities, assisted living, and affordable housing units. These gardens are organized and maintained by residents living on the premise.

- Institutional Gardens are attached to either public or private organizations and offer numerous beneficial services for residents. Benefits include mental or physical rehabilitation and therapy, as well as teaching a set of skills for job-related placement.

- Demonstration Gardens are used for educational and recreational purposes in mind. They often offer short seminars or presentations about gardening, and provide the necessary tools to operate a community garden.

Location

Community gardens may be found in neighborhoods, schools, hospitals, and on residential housing grounds. The location of a community garden is a critical factor in how often the community garden is used and who visits it. Exposure to a community garden is much more likely for an individual if they are able to walk or drive to the location, as opposed to public transportation. The length of travel time is also a factor. Those who live within a 15-minute or less travel distance are more likely to visit a community garden as compared to those with a longer travel time. Such statistics should be taken into consideration when choosing a location for a community garden for a target population.

The site location should also be considered for it's soil conditions as well as sun conditions. Solar conditions being of paramount importance, as above ground gardening is always possible. An area with fair amount of morning sunlight and shade in the afternoon is most ideal. While specifics vary from plant to plant, most do well with 6 to 8 full hours of sunlight.

When considering a location, areas near industrial zones may require soil testing for contaminants. If soil is safe, the composition should be loose and well-draining. However, if the soil at the location is unable to be used, synthetic soil may also be used in raised gardens beds or containers.

Plant Choice and Physical Layout

While food production is central to many community and allotment gardens, not all have vegetables as a main focus. Restoration of natural areas and native plant gardens are also popular, as are "art" gardens. Many gardens have several different planting elements, and combine plots with such projects as small orchards, herbs and butterfly gardens. Individual plots can become "virtual" backyards, each highly diverse, creating a "quilt" of flowers, vegetables and folk art.

Regardless of plant choice, planning out the garden layout beforehand will help avoid problems down the line. According to the Arizona Master Gardener Manual, taking measurements of the garden size, sun light locations and planted crops vs. yield quantity, will ensure a detailed record that helps when making decision for the coming years. Other consideration to garden layout would be efficient use of space by using trellises for climbing crops, being mindful of taller plants blocking sunlight to shorter plants and plants that have similar life cycles close together.

Group and Leadership Selection

The community gardening movement in North American is inclusive, diverse, pro-democracy, and supportive of community involvement. Gardeners may be of any cultural background, young or old, new gardeners or seasoned growers, rich or poor. A garden may have only a few people active, or hundreds.

Some community gardens "self-support" through membership dues, and others require a sponsor for tools, seeds, or money donations. Churches, schools, private businesses or parks and recreation departments supporters and community leaders.

Finally, all community gardens have a structure. The organization depends in part on whether the garden is "top down" or "grassroots". There are many different organizational models in use for community gardens. Some elect boards in a democratic fashion, while others can be run by appointed officials. Some are managed by non-profit organizations, such as a community gardening association, a community association, a church, or other land-owner; others by a city's recreation or parks department, a school or University.

Gardeners may form a grassroots group to initiate the garden, such as the Green Guerrillas of New York City, or a garden may be organized "top down" by a municipal agency. In Santa Clara, California there is a non-profit by the name of Appleseeds that offers free assistance in starting up new community gardens around the world.

Membership Rules and Fees

In most cases, gardeners are expected to pay annual dues to help with garden upkeep, and the organization must manage these fees. The tasks in a community garden are many, including upkeep, mulching paths, recruiting members, and fund raising. Rules and an 'operations manual' are both invaluable tools, and ideas for both are available at the ACGA.

Health Effects of Community Gardens

Community gardens have been shown to have positive health effects on those who participate in the programs, particularly in the areas of decreasing body mass index and lower rates of obesity. Studies have found that community gardens in schools have been found to improve average body mass index in children. A 2013 study found that 17% of obese or overweight children improved their body mass index over seven weeks. Specifically, 13% of the obese children achieved a lower body mass index in the overweight range, while 23% of overweight children achieved a normal body mass index. Many studies have been performed largely in low-income, Hispanic/Latino communities in the United States. In these programs, gardening lessons were accompanied by nutrition and cooking classes and optional parent engagement. Successful programs highlighted the necessity of culturally tailored programming.

There is some evidence to suggest that community gardens have a similar effect in adults. A study found that community gardeners in Utah had a lower body mass index than their non-gardening siblings and unrelated neighbors. Administrative records were used to compare body mass indexes of community gardeners to that of unrelated neighbors, siblings, and spouses. Gardeners were less likely to be overweight or obese than their neighbors, and gardeners had lower body mass indexes than their siblings. However, there was no difference in body mass index between gardeners and their spouses which may suggest that community gardening creates healthy habits for the entire household.

Participation in a community garden has been shown to increase both availability and consumption of fruits and vegetables in households. A study showed an average increase in availability of 2.55 fruits and 4.3 vegetables with participation in a community garden. It also showed that children in participating households consumed an average of two additional servings per week of fruits and 4.9 additional servings per week of vegetables.

Policy Implications

There is strong support among American adults for local and state policies and policy changes that support community gardens. A study found that 47.2% of American adults supported such policies. However, community gardens compete with the interests of developers. Community gardens are largely impacted and governed by policies at the city level. In particular, zoning laws strongly impact the possibility of community gardens. The momentum for rezoning often comes from the public need for access to fruits and vegetables. Rezoning is necessary in many cities for a parcel of land to be designated a community garden, but rezoning doesn't guarantee garden will not be developed in the future.

Further policies can be enacted to protect community gardens from future development. For example, New York State reached a settlement in 2002 which protected hundreds of community gardens which had been established by the Parks and Recreation Department GreenThumb Program from future development.

At times, zoning policy lags behind the development of community gardens. In these cases, community gardens may exist illegally. Such was the case in Detroit when hundreds of community gardens were created in abandoned spaces around the city. The city of Detroit created agricultural zones in 2013 in the middle of urban areas to legitimize the over 355 "illegal" community gardens.

Sustainable Development

Sustainable development is the organizing principle for meeting human development goals while simultaneously sustaining the ability of natural systems to provide the natural resources and ecosystem services based upon which the economy and society depend. The desired result is a state of society where living conditions and resources are used to continue to meet human needs without undermining the integrity and stability of the natural system. Sustainable development can be defined as development that meets the needs of the present without compromising the ability of future generations to meet their own needs.

While the modern concept of sustainable development is derived mostly from the 1987 Brundtland Report, it is also rooted in earlier ideas about sustainable forest management and twentieth-century environmental concerns. As the concept developed, it has shifted towards focus more on economic development, social development and environmental protection for future generations. It has been suggested that "the term 'sustainability' should be viewed as humanity's target goal of human-ecosystem equilibrium, while 'sustainable development' refers to the holistic approach and temporal processes that lead us to the end point of sustainability". Modern economies are endeavoring to reconcile ambitious economic development and obligations of preserving natural resources and ecosystems, as the two are usually seen as of conflicting nature. Instead of holding climate change commitments and other sustainability measures as a remedy to economic development, turning and leveraging them into market opportunities will do greater good. The economic development brought by such organized principles and practices in an economy is called Managed Sustainable Development (MSD).

The concept of sustainable development has been, and still is, subject to criticism, including the question of what is to be sustained in sustainable development. It has been argued that there is no such thing as a sustainable use of a non-renewable resource, since any positive rate of exploitation will eventually lead to the exhaustion of earth's finite stock; this perspective renders the Industrial Revolution as a whole unsustainable. It has also been argued that the meaning of the concept has opportunistically been stretched from 'conservation management' to 'economic development', and that the Brundtland Report promoted nothing but a business as usual strategy for world development, with an ambiguous and insubstantial concept attached as a public relations slogan.

Role of Sustainable Development in Permaculture

Permaculture is sustainable with few external inputs. Permaculture utilizes processes of evolutionary ecology to replace most of the external direct energy inputs aside from the sun, as well as all indirect inputs beyond the initial design and implementation phase. This largely cuts fossil fuels out of the system of food production.

Permaculture utilizes carefully chosen plants and domestic animals, as well as sometimes wildlife, for pest control. For example one might plant thorny hedges to keep deer out of the food production area, while providing habitat for birds of prey so that that they can prevent rodents from becoming a problem. Pests an never be entirely eliminated because predators always lag behind pests, but in this way they can be controlled so that they do not cause substantial damage in established permacultures.

A second point is that species are arranged in ways which mutually support each other. An example of an area where this might be helpful might be co-raising cattle and horses, together in an area that has snow so as to minimize the need for hay. Cattle won't eat what they can't see so they tend to starve quickly when it snows, but horses will kick the snow away to eat. Thus the horses can be used to help the cattle forage in winter. If areas are maintained with sufficient forage, hay may not even be necessary (there are recorded cases of cowboys with herds stuck in winter storms following mustang herds in order to ensure the cattle could eat, and thus saving their herds).

The Necessity for Material Lifecycles to Mimic Ecosystems

In an ecosystem, the waste product of one species or element becomes the input of another. Bears poop in the woods. Micro-organisms break down the poop and they release waste products which plants take in. Those plants may produce fruit and the bears eat the fruit. The bears then poop and the cycle repeats. This example shows the way in which nutrient cycles work in ecosystems, re-using and re-cycling all components and over time building quality soils and very efficient ways of transforming current solar energy into biological energy.

Sustainability requires, among other things, that we stop polluting our environment with things which cannot readily be re-used in the biological context. For this reason it is very important to develop closed material lifecycles. Our current factory-to-landfill must eventually be replaced with a closed cycle where very little if anything ends up in land fills because everything is re-used in some context or another. Doing this requires in essence having the output or waste of one system becoming the input of another. Recycling is one way this can happen (re-use is another), but either way waste must be re-purposed as an asset. An example might be privatizing sewage treatment, not as a waste disposal service, but as a series of cooking gas and fertilizer factories. Waste products which cannot be recycled can be re-used in unexpected ways, such as building things out of "urbanite".

The shift in thinking from a factory to landfill pipeline to a series of interconnected and infinitely continuous series of material lifecycles (akin to the water cycle, or to nutrient cycles in ecosystems), is absolutely necessary to achieving sustainability, and permaculture teaches this quite a bit. While the idea of feeding a family of 4 on a city lot is probably unrealistic, the lessons learned about sustainability and the increase in free time from doing so (instead of high maintenance flower gardens!) is well worth it.

Sustainable Communities

The emphasis on the output of one component being the input of another has the potential at least to build close-knit, sustainable communities through garden exchanges, and also things like thinking about what one can re-use things that would otherwise be discarded. Urbanite as a building material for small projects (say outdoor pizza ovens) comes to mind but the possibilities here are really endless. The more we think about what we can do with waste, the less of it ends up being discarded in unsustainable ways and the closer we become as communities as we utilize the waste products of neighbors and vice versa.

Sustainability Challenges

The big one actually is manpower (and this is an issue for most green energy too btw but for very different reasons), and the second is education. Building permacultures requires years of practice and significant training. It also turns food production into an intellectually challenging activity. However it is also less physically challenging than conventional agriculture in many cases, and so there are possibilities of closing loops here. The problem with manpower though is not so easily overlooked. With modern fossil-fuel-based agriculture it is not hard for an individual to successfully farm vast tracts of land, possibly in the hundreds of acres, with only seasonal help. Permaculture requires more people to be working in food production. It also requires that people make more careful choices regarding foods to consume (grains are less sustainable than chestnuts for example, and wheat is less sustainable than rice). More people in "farming" full-time means fewer people engineering green energy solutions and the like.

Rainwater Harvesting

Rainwater harvesting is the storing of rainwater during the monsoon season for the purpose of using it during periods of water scarcity. Generally speaking, it is a process used for collecting and storing rainwater for human use.

Rainwater harvesting is best described as the technique by which rain water is accumulated and stored with the intention of reusing it during the dry season or when there is a drought.

With rapid climatic changes, increase in global temperature and population growth, there is a scarcity of potable water in many countries across the world.

The gradual falling of water levels, are a cause of serious concern not only because it leads to shortage of usable water but also because in coastal areas it causes imbalance in salinity of the area.

Rapid industrialization and disposing chemical waste into water-bodies leads to pollution of rivers, lake and water-bodies. This is a global problem and needs a speedy solution. The supply of fresh water in this planet cannot be increased. So an alternative method must be sought for. One such method is harvesting rainwater.

Rainwater harvesting is an easy and economical way to deal with this crisis. As men are becoming environment responsible, rain water harvesting is gaining popularity leading to eco-conservation and constructive use of natural resource. Falling water tables are widespread and most people in urban areas are dependent on bottled water which is neither cost-effective nor dependable.

Importance of Rainwater Harvesting

1. Rainwater harvesting or the collection of rainwater in a proper way, can be a permanent solution to the problem of water crisis in different parts of the world. This simple method can put forward a solution which will be workable in areas where there is sufficient rain but the groundwater supply is not sufficient on the one hand and on the other surface water resource is insufficient.

This is particularly applicable in hilly areas where it can be utilized for human consumption, by animals and also for farming. In remote areas, where surface pollution is comparatively low, rainwater harvesting is ideal.

2. Although the earth is three-fourths water; very little of it is suitable for human consumption or agriculture. Rainfall is unpredictable and there is a constant shortage of water in countries which are agriculture dependent or generally drought prone.

3. A bad monsoon means low crop yield and shortage of food. Even animals suffer from scarcity of water. Africa and the Indian subcontinent face acute water crisis during the summer months. The farmers are the most affected because they do not get sufficient water for their fields. Rainwater harvesting therefore is an ideal solution for farmers who depend on monsoon for consistent water supply.

4. Unavailability of clean water compels the consumption of polluted water, giving rise to water-borne diseases and high rate of infant mortality. In recent studies it has been observed that in Lima (Peru) nearly 2 million people do not have access to any water supply and those who do have access get water supply which has a high possibility of being contaminated.

It has been reported that the water crisis in some parts of Honduras is so severe that the municipal corporation of those areas cannot supply enough water even to those residents who have municipal water supply connection. This has been reported by Anna Kajumuto Tibaijuka, Under-secretary General, UNED UN-Habitat.

5. If rain water, which comes for free, can be collected and stored, instead of letting it run off, it could be an alternative to back up the main water supply especially during dry spells. Its importance will not be limited to an individual family but can be used by a community as well.

Systematic rainwater harvesting can help in irrigation with minimum use of technology and is therefore cost effective. This simple method can help farmers to prevent their crops from drying due to lack of water. It also creates a sense of social responsibility and awareness about the environment.

6. The importance of rainwater harvesting lies in the fact that it can be stored for future use. Just as it can be used directly so also the stored water can be utilized to revitalize the ground level water and improve its quality. This also helps to raise the level of ground water which then can be easily accessible. When fed into the ground level wells and tube well are prevented from drying up. This increases soil fertility. Harvesting rainwater checks surface run off of water and reduces soil erosion.

7. In areas having sparse and irregular rainfall, scarcity of water is a persistent problem. It cannot be completely resolved but can be mitigated through rain water harvesting. Rainwater harvesting is an ideal solution to water problems in regions which receive inconsistent rainfall throughout the year.

Methods of Rainwater Harvesting

The most common methods for rainwater harvesting are:

1.Surface run off harvesting: Surface run off harvesting is most suitable in urban areas. Here rain water flows away as surface run off and can be stored for future use. Surface runoff rain water in ponds, tanks and reservoirs built for this purpose. This can provide water for farming, for cattle and also for general domestic use. Without sufficient water health and hygiene are severely affected. This adds to the environmental pollution.

Surface water can be stored by redirecting the flow of small creeks and streams into reservoirs on the surface or underground.

2. Roof top rain water harvesting: Roof top water harvesting can be done in individual homes or in schools.

For this the first requirement is to intercept the rainwater to flow towards a definite direction.

The water should reach a bucket or a tank through pipes made from wood or bamboo. In urban areas pvc pipes can be used.

The rain water can be directly collected by keeping a bucket or container beneath the roof.

The first flow of rain water will usually carry with it dust particles, leaves, insects and bird droppings. So it is best to use a detachable downpipe to divert the first rain water.

Recharge pits can be constructed to hold rain water. These can be of any shape or size, depending upon the amount of rain the area receives. These pits need to be filled with, boulders, gravels and coarse sand, which will act as a filter for the impurities that are carried along with the first flow of water.

Roof top rain water can be harvested through existing tube wells. In areas where the aquifer that holds the ground water has dried up, tubewells source deeper into the soil for water. Roof top rain water harvesting can be done through these dried up tubewells to rehydrate the dried subsoil water level.

As man became more and more dependent on technology, he moved away from nature and perhaps ignored the value of natural resources. In his march towards power, he forgot to save the precious water which is another name for life. Probably because most of the natural resources come for free we fail to how precious it is until a worldwide crisis was issued.

It is our unfriendly attitude towards nature that has polluted water-bodies and left them unfit for any use. Natural resources come in abundance but they cannot be produced in our workshops. Today most of the countries of the world are facing the scarcity of water and are taking up rainwater harvesting for the dire necessity of survival. So the seriousness of the issue has been realized the seriousness and efforts to overcome this problem has been initiated.

Rainwater Harvesting for Sustainable Agriculture

Water is an important resource that is used in our daily lives. It used in vitally important sectors of the economy, such as the agriculture sector. Farmers use water to grow crops. Not only is water used to grow crops, it is also used to process agricultural products before they can be sent to the marketplace. Even when they reach markets and are bought by consumers, water is still needed to transform raw food items into edible forms. Water is indisputable an essential resource used by everyone linked to the agriculture sector.

Water conservation is increasingly being encouraged in crucial sectors of the economy, such as the agriculture sector. This is fuelled by an increasing demand for water and growing concerns of water scarcity in the society. The United Nations even considers water availability to be a major issue for the 21st century.

Rainwater harvesting has agricultural uses. It can be used for watering gardens in our homes and crop plants in agricultural fields. These reduce the reliance of garden owners and farmers on other sources of water supply, thus saving them money.

Also, we are in a climate change era where intense rainfall is expected. And it can damage agricultural land areas. Rainwater harvesting can be used to divert heavy rainfall from reaching agricultural lands, thereby protecting crop plants from getting damaged.

One good property of rainwater is that it is a soft form of water and does not impact plants negatively. Unlike hard water, that adds calcium carbonate to crop plants, forming a coating on the roots/leaves. When such coatings are formed, it prevents plants from receiving the maximum amount of the water, minerals, fertilizers and pesticides that are supplied to them. It also prevents plants from receiving maximum sunlight, thereby slowing down photosynthesis.

Furthermore, the use of soft water from rainwater harvesting can help to reduce farmers operating costs. This is because calcium carbonate from hard water normally piles up in pumps or sprinklers causing blockages. When such equipment is blocked, money is used to unblock their pathways. In contrast, such problems are not usually associated with the use of soft water in farming operations, thus reducing the cost to maintain crops.

More so, the use of hard water from water mains in farming operations causes scale formations on plants- due to the Calcium Carbonate contents of hard water. These formations promote the growth of bacteria that is capable of damaging crop plants. However, such scale formations are not linked with the use of soft water from rainwater in farming operations, make it safe for plants.

Also, rainwater can be used as a source drinking water for livestock. And it is suitable for livestock compared to chlorinated water.

Furthermore, rainwater can be used to carry out domestic tasks in the farm such as cleaning machinery.

References

- Organic-farming-benefits: conserve-energy-future.com, Retrieved 13 July, 2019

- What-is-the-relationship-between-permaculture-and-sustainability-595: sustainability.stackexchange.com, Retrieved 14 August, 2019

- Ferris, J.; Norman, C.; Sempik, J. (2001). "People, Land and Sustainability: Community Gardens and the Social Dimension of Sustainable Development". Social Policy and Administration. 35 (5): 559–568. doi:10.1111/1467-9515.t01-1-00253

- Importance-methods-rainwater-harvesting: eartheclipse.com, Retrieved 05 January, 2019

- Rainwater-harvesting-sustainable-agriculture: permaculturenews.org, Retrieved 16 April, 2019

- Atkinson, G., S. Dietz, and E. Neumayer (2009). Handbook of Sustainable Development. Edward Elgar Publishing, ISBN 1848444729

Environmental Design

7

- **Agroecology**
- **Landscape Design and Planning**
- **Layers of Forest**

A design that aims to improve and enhance natural, social, cultural and physical aspects of the surroundings of a specific area is termed as environmental design. Agroecology, landscape design and planning and layers of forest such as canopy, understory, shrub layer, etc. are among its components. All the diverse components related to environmental design have been carefully analyzed in this chapter.

Permaculture is a lifestyle that encompasses ethics and design in order to mimic the relationships seen in nature. The goal is to view every environment as one part of the whole world. Rather than being consumers, the focus is on becoming "responsible producers" of biodiversity and returning any unused goods to the ground. Permaculture aims to build long-standing cultures that can sustain or be sustained for extended periods of time. This path focuses on food forest development with more permanent, perennial plant life. Permaculture, or Linking Science, offers plausible solutions for many environmental issues that exist today due to toxic human behavior.

The impact that humans and our food culture have had on the natural environment is disgraceful, and the harsh effects of intensive agriculture continue toxifying the environment. Many of our current environmental problems are the result of excessive herbicides, pesticides, and chemical fertilizers. All are common approaches to address pest or chemical issues without conceptualizing the repercussions of these actions on our resources.

"Soil erosion, ocean acidification, atmospheric pollution, bioaccumulation of toxic substances, declining sources of drinkable fresh water, deforestation, waste buildup, diminishing biodiversity, species extinction, peak oil, climate change, and valuable resource depletion all threaten the integrity of the vastly complex and fragile society/biosphere relationship". As stewards of this planet, it is our responsibility to start looking at life in a more holistic manner and treating plants, ourselves, and the planet generally as part of a living ecosystem.

Permaculture creates a circular, holistic ecosystem in which everything is created from the Earth in order to care for other life. Humans give back surplus items to the Earth to be reused as viable sources of energy for another life form, just as animals and plants would. When it comes to food production, permaculture aims to create bio-diverse food forests. The unsustainable monocultural systems that are commonly practiced throughout the world don't offer such vast options as food forestry through proper permaculture techniques.

One species, humans, is taking 40% of the biological productivity of the planet's entire land area, mostly to provide well for only 1 billion people. If another 10 billion want to live as we in the rich countries do how much habitat will be left for the other possibly 30 million species? We cannot possibly expect to stop the extinction of species unless we drastically reverse this demand for biological resources and the consequent destruction of habitat.

Food forests alleviate the need for synthetic intervention so long as they are well structured. They also provide enough food for humans and animals to graze and collect surplus while using much less land. Substances that cannot biodegrade are never put into the soil or water which limits outgoing waste from communities.

Permaculture has been gaining popularity and interest since it was first developed in 1978 by Australian Bill Mollison. It has since spread over the entire world, as farmers and families try to control their survival necessities and reduce the human footprint. There are many recognized permaculture schools, and courses are beginning to show up in colleges and online, sometimes for free. Most people in the permaculture field are well networked, as making connections is a necessary part of the practice. This circular process is necessary in all relationships of life, including everything from human relations to caring for animals to growing food and processing waste materials from humans, plants, and animals.

The most developed network that currently exists for permaculture students and practitioners is called the Worldwide Permaculture Network. With this tool, people are able to add their projects to a global map with descriptions of their needs and desires. Thanks to the Permaculture Research Institute of Australia, the networking power available surpasses many other movements that have similar goals of a greener planet. Perhaps it is because of the way permaculture is taught that the people involved know that "many hands make light work".

Humans have been struggling to coexist within the planet's environmental systems for quite some time, but never before in history have we been in such a predicament. We can create a system that is interconnected and self reliant but this system, in order to gain full benefits, will not work in our current consumer society. "Permaculture can very easily be part of the problem. It is part of the problem if does not increase the realisation that affluent living standards and this economy are totally incompatible with sustainability and with global economic justice".

If we want a more sustainable world, we have to change by reducing consumption and increasing sustainable production. The planet cannot sustain life at the rate it is currently. Environmental problems will only get worse if we don't take direct action. Permaculture offers an approach to sustainability that is hard to find, but it also requires determination beyond simply educating others.

The structure of our global food chain is appalling, and it is up to everyone who cares to make a statement about the kind of world we want. How we go about this is the determining factor in whether we can really create a better life for ourselves and the planet. It is saddening to think about the damage we have caused to this bountiful place we call Earth, but it doesn't have to be this way.

To prevail, we as consumers can take strong action, making statements about the world we want to support by simple restrictions. For example, where we spend our money makes a huge statement about what is important to us. How much interest we show toward something makes it either worthwhile or not, and this can drive up the monetary value of products. If we put our money where our heart is and our hands in the soil, we can create a more beautiful world with less corruption, poverty, and detrimental environmental impacts.

Agroecology

Increasing concerns about the negative impacts of industrial agriculture have generated a vigorous debate over the feasibility of transition to alternative forms of agriculture, capable of providing a broad suite of ecosystem services while producing yields for human use. The transition to diversified, ecologically benign, smaller scale production systems is addressed in the literature of agroecology, diversified farming systems, and multifunctional agriculture. Agroecological transition must be regarded as a complex, multi-sector project, operating at multiple temporal and spatial scales and involving diverse constituencies. For this reason, researchers have often directed their attention outside of institutional science to document the contributions that traditional and innovative practices offer to the process of transition. Alternative agroecology movements, for example, have been critical in the process of regional agroecological transition and likely will be in the future.

Permaculture is an international movement and ecological design system. Despite permaculture's international extent and relatively high public profile, it has received very little discussion in the scientific literature. The term originated as a portmanteau of permanent agriculture and is defined by co-originator David Holmgren as "Consciously designed landscapes which mimic the patterns and relationships found in nature, while yielding an abundance of food, fibre and energy for provision of local needs". As a broadly distributed movement with a distinctive conceptual framework for agroecosystem design, permaculture's relevance to the project of agroecological transition has several aspects. Permaculture can function as a framework for integrating knowledge and practice across disciplines to support collaboration with mixed groups of researchers, stakeholders, and land users. Permaculture contributes to an applied form of ecological literacy, supplying a popular and accessible synthesis of complex socioecological concepts. The design orientation of permaculture offers a distinctive perspective that suggests avenues of inquiry in agroecosystem research. Lastly, these factors are embodied in an international movement that operates largely outside of the influence and support of large institutions, which suggests opportunities for participatory action research and the mobilization of popular inquiry and support.

The potential of permaculture to contribute broadly to agro-ecological transition is limited by several factors. Of primary importance is the general isolation of permaculture from science, both in terms of a lack of scholarly research about permaculture and neglect within the permaculture literature of contemporary scientific perspectives. This deficit is compounded by overreaching and oversimplifying claims made by movement adherents and the absence of any systematic multisite assessment of permaculture's impacts. Additionally, the difficulty of providing a clear and distinguishing description of perma-culture can cause confusion and hinder rigorous and systematic discussion.

Introductory material includes a brief overview of the origins and development of permaculture, the growth of the movement over time, and a preliminary heuristic for comparing the prominence and overlap of permaculture and agro-ecology across several sectors. The introduction is followed by a systematic review of scientific and popular permaculture literature, analyzing publication type, date, and location, topic location, scholarly discipline, and citations. Systematic analysis also includes quantitative content analysis using a concept.

Examples of production and education in the permaculture movement. a Small farm with inter-cropped annuals and perennials, worked partially with hand labor. b Workshop on the design and maintenance of perennial polycultures network approach. Qualitative review of the permaculture framework then identifies and evaluates prominent themes in the permaculture literature, focusing on agroecological topics. Finally, qualitative and quantitative analyses are synthesized to produce an overall evaluation of permaculture, including recommendations for future directions for research and dialog.

Conceptual Influences

Permanent as Sustainable and Perennial

The term permanent agriculture, from which the word perma-culture is derived, has multiple uses. Permanent agriculture is used to contrast sedentary, continuous agriculture with shifting cultivation in discussions of the latter. Examination of the British and US literature on farming practices in the early 1900s suggests that the word "permanent" was used in an analogous fashion to the current use of the term sustainable. With the publication of Russell Smith's foundational agroforestry text Tree Crops: A Permanent Agriculture, permanent came to connote agricultural systems incorporating a high proportion of perennial species. It is this concept for which permaculture is named. Mollison and Holmgren adopted Smith's emphasis on the importance of tree crops for soil stabilization in hillside agriculture, production of fodder, and production of complementary and staple foods for human consumption. The portmanteau of "permanent agriculture" was later redefined as "permanent culture" as the scope of permaculture expanded from the design of small-holder agriculture to encompass human settlement more broadly.

Systems Ecology

Permaculture's emphasis on whole systems design is heavily influenced by the work of ecologist H.T. Odum. Odum developed the influential framework of systems ecology, a thermodynamic perspective that regards ecosystems as networks through which energy flows and is stored andtrans-formed, which can be diagramed and modeled in a manner analogous to electronic circuits. Odum referred to the applied form of systems ecology as ecological engineering, and this design perspective would shape fundamental components of the permaculture perspective. In the highly cited book Environment, Power, and Society, Odum proposes an approach to the design of novel and productive ecosystems in which species are regarded as distinctive but interchangeable system components which should be selected from a global pool without regard to the place of origin. In this view, the distinctive inputs and outputs of each species will connect in novel assemblages, and the exchanges of energy and resources between system components will substitute for human labor and material inputs. Ecosystem designers should therefore foster self-organization through the iterative "seeding" of diverse species from the global species pool, in order to generate and select ecosystems which produce yields for human use with minimal labor input. The influence of this focus on functional relationships between components, the self-organization of systems, and species selection practices is reflected throughout the permaculture literature.

Keyline Planning

Holmgren and Mollison were also informed by the whole landscape approach of the Australian Keyline design system. From the 1950s to the 1970s, farmer and writer P.A. Yeomans developed a system that integrated novel methods for landscape analysis with whole farm water management, agroforestry, soil building strategies (using slightly-off-contour chisel plowing and rotational grazing), and the development of new chisel plow designs for use in the system. Yeoman's Keyline system has received very little attention in the scientific literature. Keyline planning is nevertheless an innovative application of design to agricultural landscapes and shaped the approach taken by Holmgren and Mollison, who adopted many of the concepts of the Keyline plan directly into the developing permaculture framework.

Permaculture and Agroecology

In the past three decades, permaculture has grown in parallel with agroecology, displaying overlapping concerns while developing different constituencies. Permaculture shares with the discipline of agroecology a focus on the intersection of ecology and agricultural production, a normative orientation toward agroecological transition, and an association with popular movements consisting largely of land users. Despite these parallels, permaculture has received very little discussion in the agroecological literature. When permaculture is mentioned, it is frequently found as an item on a list of alternative agricultural frameworks, the value of which is either explicitly in question, or positive but nonspecific. Permaculture is elsewhere associated positively, albeit in passing, with agroforestry, perennial polycultures, agroecosystem design, ecosystem mimicry, and agrobiodiversity. Substantive assessment of permaculture as an approach to agriculture, positive and negative, appears to be absent from the peer-reviewed literature.

This absence is surprising in light of permaculture's international public profile. Parallel queries of online databases for the terms "permaculture" and "agroecology" can be used to illustrate patterns in the relative prominence and overlap of each field across sectors. This fairly crude comparison is presented here in a preliminary fashion to demonstrate that the sparse representation of permaculture in the scientific literature is incommensurate with a high level of general interest. The proportions of results returned for each term varied widely across data sources. The scientific databases Web of Knowledge and Google Scholar returned 21 and 6 times as many results for agroecology as for permaculture, respectively, while general purpose internet search engines Google and Bing were skewed in the opposite direction, returning 11 and 7 times as many results for permaculture as for agroecology, respectively. Multipurpose literature databases for book sales were less asymmetrical, with approximately equal results for each term in Google Books and twice the results for permaculture in Amazon. Document archives of international development organizations (US AID, Peace Corps, and FAO) were highly and heterogeneously skewed, respectively, returning 3 times the results for agroecology as for permaculture, 41 times the results for permaculture, and 21 times the results for agroecology.

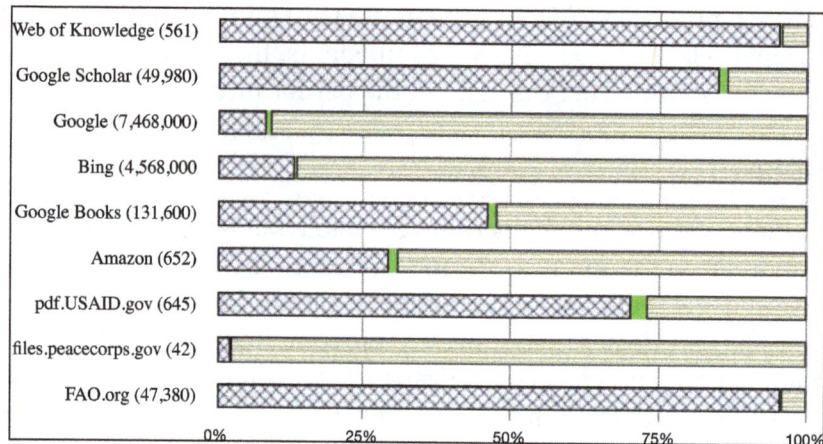

Proportional results from parallel search queries for "agroecology" (crosshatch), combined "agroecology" + "permaculture" (solid), and "permaculture" (horizontals), to multiple online data sources, illustrating the uneven relative prominence of agroecology and permaculture across different sectors. Numbers in parentheses indicate combined total responses from each data source.

In addition to the parallels, permaculture shares with agroecology a complex stratified definition. Recent scholarship has clarified that agroecology simultaneously refers to a scientific discipline, a social movement, and a set of agricultural practices. Similarly, some of the confusion surrounding permaculture may be attributed to the use of the term to refer to a design system, to an international movement, to the worldview carried by and disseminated by the movement, and to the set of associated practices. Figure is a conceptual map intended to clarify the relationship among the different strata that make up permaculture, each of which intersects with the project of agroecological transition. This conceptual structure will be used to organize the examination and assessment of the permaculture literature.

Landscape Design and Planning

As more and more people learn about the problems that industrial agriculture (and industrial civilization in general) have caused the world, there is a growing interest in all things organic. From larger organic sections at your grocery store to the explosion of the local food movement with farmer's markets and community supported agriculture programs, organic livelihoods are on the rise.

When it comes to actually growing your own food in a sustainable and ecologically healthy manner, however, it takes much more than just a little bit of compost. The importance of learning how to read the landscape and discover what elements belong and can fit into the overall functioning of the system is an essential part of sustainable living on the land. Permaculture is one of the most helpful tools to help us learn how to best design our livelihoods to the land.

Our modern-day society has very little consideration for design, at least any sort of design that takes into account the realities of the natural world. We have grown to believe that as humans, our ultimate purpose is to bring the natural world under our control and make it work for us. The natural world, when we do see it, is relegated to "wilderness" areas that can be found in national parks. We vacation to see the world as it is, but the rest of the time want the world to conform to our needs and desires. Some people blame this separation between the human and his or her natural surrounding on the Industrial Revolution and our new-found ability to use fossil fuels to help along our inner desire for growth and dominion over the natural world.

As technology has developed over time, we have increasingly lost contact with the natural rhythms of the natural world. If you ask any stranger street what mood we are in, they will probably give you a quizzical look. Most elementary school students will be able to identify hundreds of corporate logos and brands, but can´t tell the difference between an oak tree and an elm tree. Technology has isolated us from the world around us in two ways. Firstly, it removes us from any sort of proximity with the natural world. We wake up in the morning not to the sound of birds, but to an alarm clock playing our chosen music. Though it may be snowing outside, we walk around our house barefoot and in pajamas thanks to the wonders of central heating. Afraid to brave the cold for even a moment, we push a button from inside our house to start up our car and get it heated before we begin our drive to work.

Secondly, technology has effectively erased the ebb and flow of the seasons. A trip to the local grocery store won´t show you in-season and out of season produce. We can get fresh strawberries in the dead of winter and enjoy bananas year-round thanks to a food system dependent on long transportation and cheap wage labor in Mexico and around the world. Permaculture is an attempt to put the natural world back into the equation, to understand that any human economy exists within a larger economy that includes the birds, the soil, the rain, and the aquifers. In essence, the process of permacultural design is to help humanity rediscover the original meaning of economy (oikonomia) as the "ordering of the household". Our household includes the place we live, the creatures we share it with, and the balance and equilibrium that our lives must conform to if we are to continue to exist as a species that shares this only world of ours.

Nature as Measure

Permaculture design begins with simply being present in place and slowly getting a feel for what that place will allow one to do. Whereas most "development" in modern day society takes place behind an architect's desk or in the mind of some agronomist, permaculture urges people to begin by actually spending time in the place where you live. Observing the rhythms of the natural world that we have been taught to forget and seeing patterns emerge from the natural functioning of the world itself are the most important parts of any ecological design. This only happens by accepting a physical proximity and intimacy with our places and learning that Nature should be the measure of our success.

As we relearn these patterns and rhythms, it should become obvious to us that our lives and livelihoods need to conform to the natural functioning of the place. No longer should we consider ourselves to be the epitome of evolution with a divine right to impose our will wherever we go, but rather learn to limit aspects of our livelihood. The idea of finding freedom and belonging through natural limitations is almost a heresy. People who do make the decision to live simply and respectfully of their places are seen to be "wasting" their lives and their talents. For people who do attempt to discover the simplicity of living within natural limits, however, they also find natural opportunities that are many times hidden from the sight of people whose vision is always on the horizon and not on the ground beneath them.

The Science of Design: An Example

On a practical level, living close to the land can help one discover the opportunities that the modern, industrial mind simply cannot see. In rural Kentucky, the son of a wealthy businessman inherited five acres of forest in the foothills of the Appalachian Mountains. His father had kept the land as a reminder of his humble beginnings, but his son could only see it as an antiquated place of backwardness and poverty. As soon as he could, he sold the land and invested his time and money in other ventures.

The small farmer who bought the forest was considering clearing the land to grow tobacco, but while wandering through the forest, he found morel mushrooms growing wild throughout the forest floor. He invested a small amount of money in mushroom spawn, thinned the forest for logs, and began an extremely profitable gourmet mushroom business that maintained the forest ecosystem healthily intact while at the same time providing for himself and his family.

The small farmer's observation of the natural world and his ability to listen to what the forest offered helped him design a livelihood that respected the boundaries and limits of that place while also reaping the benefits of the opportunities that were naturally present. Had he blindly followed his initial anthropocentric idea of knocking over the forest to grow tobacco, the forest ecosystem would have been lost and he almost certainly would have encountered financial ruin as the tobacco industry plummeted shortly after his purchase of the forest.

The Importance of Design

Of the many different contributions that the permaculture movement has offered to folks interested in living more sustainably, the focus on permaculture as a design process is perhaps the most relevant. Learning to observe the natural world, accept Nature as a measure with the limitations and opportunities that are inherently present in each and every place, and design our livelihoods according to this reality is an indispensable part of living correctly on the land.

Layers of Forest

Canopy

In biology, the canopy is the aboveground portion of a plant community or crop, formed by the collection of individual plant crowns.

In forest ecology, canopy also refers to the upper layer or habitat zone, formed by mature tree crowns and including other biological organisms (epiphytes, lianas, arboreal animals, etc.).

Sometimes the term canopy is used to refer to the extent of the outer layer of leaves of an individual tree or group of trees. Shade trees normally have a dense canopy that blocks light from lower growing plants.

The canopy of a forest in Sabah, Malaysia.

Canopy Structure

The canopy of a forest.

Bamboo canopy in the Western Ghats of India.

A monkey-ladder vine canopy over a road.

Canopy structure is the organization or spatial arrangement (three-dimensional geometry) of a plant canopy. Leaf area index (LAI), leaf area per unit ground area, is a key measure used to understand and compare plant canopies. It is also taller than the understory layer. The canopy holds 90% of the animals in the rainforest. They cover vast distances and appear to be unbroken when observed from an airplane. However, despite overlapping tree branches, rainforest canopy trees rarely touch each other. Rather, they are usually separated by a few feet.

Canopy Layer of Forests

Dominant and co-dominant canopy trees form the uneven canopy layer. Canopy trees are able to photosynthesize relatively rapidly due to abundant light, so it supports the majority of primary productivity in forests. The canopy layer provides protection from strong winds and storms, while also intercepting sunlight and precipitation, leading to a relatively sparsely vegetated understory layer.

Forest canopies are home to unique flora and fauna not found in other layers of forests. The highest terrestrial biodiversity resides in the canopy of tropical rainforests. Many rainforest animals have evolved to live solely in the canopy, and never touch the ground.

The canopy of a rainforest is typically about 10m thick, and intercepts around 95% of sunlight. The canopy is below the emergent layer, a sparse layer of very tall trees, typically one or two per hectare. With an abundance of water and a near ideal temperature in rainforests, light and nutrients are two factors that limit tree growth from the understory to the canopy.

In the permaculture and forest gardening community, the canopy is the highest of seven layers.

Understory

In forestry and ecology, understory (or understorey, underbrush, undergrowth) comprises plant life growing beneath the forest canopy without penetrating it to any great extent, but above the forest floor. Only a small percentage of light penetrates the canopy so understory vegetation is generally shade tolerant. The understory typically consists of trees stunted through lack of light, other small trees with low light requirements, saplings, shrubs, vines and undergrowth. Small trees such as holly and dogwood are understory specialists.

In temperate deciduous forests, many understory plants start into growth earlier than the canopy trees to make use of the greater availability of light at this time of year. A gap in the canopy caused by the death of a tree stimulates the potential emergent trees into competitive growth as they grow upwards to fill the gap. These trees tend to have straight trunks and few lower branches. At the same time, the bushes, undergrowth and plant life on the forest floor become more dense. The understory experiences greater humidity than the canopy, and the shaded ground does not vary in temperature as much as open ground. This causes a proliferation of ferns, mosses and fungi and encourages nutrient recycling, which provides favorable habitats for many animals and plants.

Lesser celandine (Ranunculus ficaria) on forest floor in spring.

Understory Structure

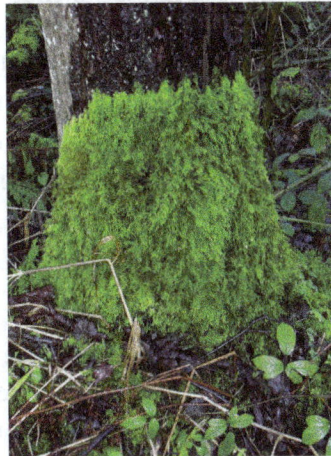

Tree base showing moss understory limit.

The understory is the underlying layer of vegetation in a forest or wooded area, especially the trees and shrubs growing between the forest canopy and the forest floor. Plants in the understory comprise an assortment of seedlings and saplings of canopy trees together with specialist understory shrubs and herbs. Young canopy trees often persist in the understory for decades as suppressed juveniles until an opening in the forest overstory permits their growth into the canopy. In contrast understory shrubs complete their life cycles in the shade of the forest canopy. Some smaller tree species, such as dogwood and holly, rarely grow tall and generally are understory trees.

The canopy of a rainforest is typically about 10m (33ft) thick, and intercepts around 95% of the sunlight. The understory receive less intense light than plants in the canopy and such light as does penetrate is impoverished in wavelengths of light that are most effective for photosynthesis. Understory plants therefore must be shade tolerant—they must be able to photosynthesize adequately using such light as does reach their leaves. They often are able to use wavelengths that canopy plants cannot. In temperate deciduous forests towards the end of the leafless season, understory plants take advantage of the shelter of the still leafless canopy plants to "leaf out" before the canopy trees do. This is important because it provides the understory plants with a window in which to photosynthesize without the canopy shading them. This brief period (usually 1–2 weeks) is often a crucial period in which the plant can maintain a net positive carbon balance over the course of the year.

As a rule forest understories also experience higher humidity than exposed areas. The forest canopy reduces solar radiation, so the ground does not heat up or cool down as rapidly as open ground. Consequently, the understory dries out more slowly than more exposed areas do. The greater humidity encourages epiphytes such as ferns and mosses, and allows fungi and other decomposers to flourish. This drives nutrient cycling, and provides favorable microclimates for many animals and plants, such as the pygmy marmoset.

Shrub Layer

The shrub layer is the stratum of vegetation within a habitat with heights of about 1.5 to 5 metres. This layer consists mostly of young trees and bushes, and it may be divided into the first and second shrub layers (low and high bushes). The shrub layer needs sun and little moisture, unlike the moss layer which requires a lot of water. The shrub layer only receives light filtered by the canopy, i.e. it is preferred by semi-shade or shade-loving plants that would not tolerate bright sunlight. Small to medium sized birds sometimes known as *bush nesters* are often found in the shrub layer where their nests are protected by foliage. European examples include blackbird, song thrush, robin or blackcap. In addition to shrubs, such as elder, hazel, hawthorn, raspberry and blackberry, clematis may also occur while, in other parts of the world, vines and lianas may form part of this stratum. At the edge of a woodland the shrub layer acts as a windbreak close to the trees and protects the soil from drying out.

Herbaceous Layer

The herbaceous stratum contains non-woody vegetation, or ground cover, growing in the forest with heights of up to about one and a half metres. The herbaceous layer consists of various herbaceous plants, grasses, dwarf shrubs (hemicryptophytes, geophytes, therophytes and chamaephytes) and young shrubs. In forests, early flowering plants appear first before the canopy fills out. Thereafter, the amount of light available to plants is significantly reduced and only those that are suited to such conditions can thrive. By contrast, grassland consists of moss and herbaceous layers. Sometimes, a shrub layer builds up as part of a process of reforestation (succession).

Groundcover

Groundcover or ground cover is any plant that grows over an area of ground. Groundcover provides protection of the topsoil from erosion and drought.

In an ecosystem, the ground cover forms the layer of vegetation below the shrub layer known as the herbaceous layer. The most widespread ground covers are grasses of various types.

In ecology, groundcover is a difficult subject to address because it is known by several different names and is classified in several different ways. The term groundcover could also be referring to "the herbaceous layer," "regenerative layer", "ground flora" or even "step over".

In agriculture, ground cover refers to anything that lies on top of the soil and protects it from erosion and inhibits weeds. It can be anything from a low layer of grasses to a plastic material. The term *ground cover* can also specifically refer to landscaping fabric which is like a breathable tarp that allows water and gas exchange.

In gardening jargon, however, the term *groundcover* refers to plants that are used in place of weeds and improves appearance by concealing bare earth.

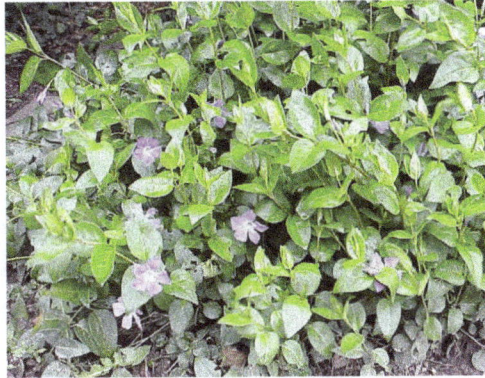
Groundcover of Vinca major.

Contributions to the Environment

The herbaceous layer is often overlooked in most ecological analyses because it is so common and contributes the smallest amount of the environment's overall biomass. However, groundcover is crucial to the survival of many environments. The groundcover layer of a forest can contribute up to 90% of the ecosystem's plant diversity. Additionally, the herbaceous layer ratio of biomass to contribution to plant productivity is disproportionate in many ecosystems. The herbaceous layer can constitute up to 4% of the overall net primary productivity (NPP) of an ecosystem, four times its average biomass.

Reproduction

Groundcover typically reproduces one of five ways:

- Lateral growth.
- Side growth: Branches on the side of the plant extend outwards upon contact with the soil.
- Base growth: New plants produced from the base of the origin plant.
- Under/Above Ground growth: Produced from rhizomes and stolons.
- Roots.

Like most foliage, groundcover reacts to both natural and anthropogenic disturbances. These responses can be classified as legacy or active responses. Legacy responses occur during long-term changes to an environment, such as the conversion of a forest to agricultural land and back into forest. Active responses occur with sudden disturbances to the environment, such as tornadoes and forest fires.

Groundcover has also been known to influence the placement and growth of tree seedlings. All tree seedlings must first fall from their origin trees and then permeate the layer created by groundcover in order to reach the soil and germinate. The groundcover filters out a large amount seeds, but lets a smaller portion of seeds pass through and grow. This filtration provides ample amount of space between the seeds for future growth. In some areas, the groundcover can become so dense that no seeds can permeate the surface, and the forest is instead converted to shrubbery. Groundcover also inhibits the amount of light which reaches the floor of an ecosystem. An experiment conducted with the *rhododendron maximum* canopy in the southern Appalachian region concluded that 4 to 8% of total sunlight makes it to the herbaceous layer, whereas only about 1 to 2% reaches the ground.

Variation

Two common variations of groundcover are residency and transient species. Residency species typically reach a maximum of 1.5 meters in height, and are therefore permanently classified as herbaceous. Transient species are capable of growing past 1.5 meters, and are therefore only temporarily considered herbaceous. These height differences make ideal environments for a variety of animals, such as the reed warbler, the harvest mouse and the wren.

Groundcover can also be classified in terms of its foliage. Groundcover that keeps its foliage for the entire year is known as evergreen, whereas groundcover that loses its foliage in the winter months is known as deciduous.

In Gardening

Microstegium vimineum, an invasive groundcover.

Five general types of plants are commonly used as groundcovers in gardening:

- Vines, which are woody plants with slender, spreading stems.

- Herbaceous plants, or non-woody plants.

- Shrubs of low-growing, spreading species.

- Moss of larger, coarser species.

- Ornamental grasses, especially low-growing varieties.

Of these types, some of the most common groundcovers include:

- Alfalfa (*Medicago sativa*).
- Clover (*Trifolium*).
- Dichondra.
- Bacopa (*Bacopa*).
- Ivy (*Hedera*).
- Gazania (*Gazania rigens*).
- Ground-elder (*Aegopodium podagraria*).
- Ice plant.
- Japanese honeysuckle (*Lonicera japonica*).
- Junipers of various low-growing types.
- Lantana, creeping species.
- Lilyturf (*Liriope muscari* and *Liriope spicata*).
- Mint (*Mentha*).
- Nasturtium (*Tropaeolum majus*).
- Pachysandra.
- Pearlwort (*Sagina subulata*).
- Periwinkle (*Vinca*).
- Shasta daisy (*Leucanthemum*).
- Soleirolia (*Soleirolia soleirolii*).
- Spider plant (*Chlorophytum comosum*).

In Roof Gardens

Groundcover is a popular solution for difficult gardening issues because it is low maintenance, aesthetically pleasing and fast growing, minimizing the spread of weeds. For this reason, ground cover is also a common choice for roof gardens. Roofs take on the brunt of incoming weather, meaning any plants on a roof must be resistant to long-term exposure to sun, overwatering from rain and harsh winds. Groundcover plants are able to sustain themselves in such conditions while also providing lush vegetation to what would otherwise be unused space.

Rhizosphere

The rhizosphere is the word used to describe the area of soil surrounding plant roots. It is the most biologically active layer of the soil; populated with micro organisms interacting and benefiting from chemicals released by plant roots. There are more micro organisms present in a teaspoon of soil than there are people on the earth; the rhizosphere can carry 1000-2000 times this amount making it highly populated with microbial life. The rhizosphere has three zones the endorhizosphere, the rhizoplane and the ectorhizosphere.

- The endorhizosphere: This is the inner section of the rhizosphere. It is the section of the plant root occupied by micro organisms (which benefit from organic compounds released by roots).

- The rhizoplane: This is the middle section of the rhizosphere. It includes the plant root surface with soil particles adhering to it.

- The ectorhizosphere: The outer area of soil surrounding the roots.

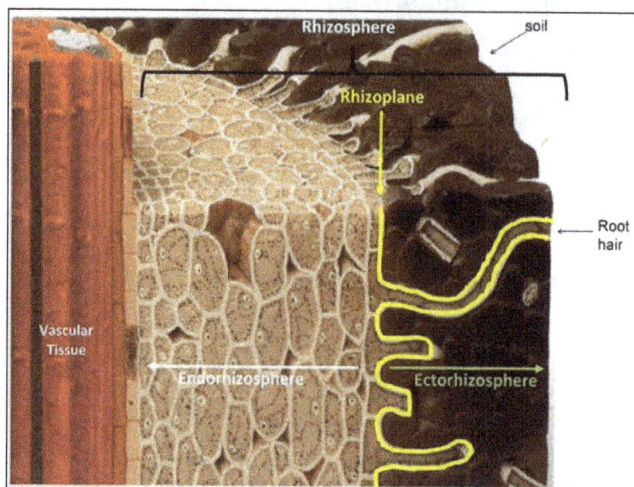

Image of the rhizosphere showing its three described sections-the endorhizosphere, the rhizoplane and the ectorhizosphere.

Rhizodeposition

Rhizodeposition is what makes the rhizosphere an interesting place for biological study. Plant roots release organic compounds into their surrounding soils. These compounds are what scientists refer to as rhizodeposits and the process is called rhizodeposition. As plants release organic compounds from their roots they lose some of their carbon to the surrounding soil; this means that they contribute to the soil carbon content. So the soil benefits from the plants lost carbon. The release of organic compounds in the root zone also supports the soil microbial life. The amount and composition of rhizodeposits from plant roots vary depending on the type of plant, the climate, the nutrient deficiency, and the physical, chemical and biological properties of soil surrounding the root. Rhizodeposits are released through the following ways:

- Production of mucilage: Mucilage is a thick, insoluble, polysaccharide rich substance produced by the cells of the root cap. Their function is to lubricate and protect the root tips as they penetrate through the soil. Mucilage is produced in many plants and also has the ability to improve the soil structure by binding together soil particles to form aggregates.

- The release of root exudates: Root exudates are released by plant roots. Scientific research has shown that plants release exudates for a number of reasons (a) to restrict the growth of competing plant species (b) to attract symbiotic relationships (for example, the legume-rhizobia relationship which helps to fix atmospheric nitrogen) (C) to change the chemical or physical properties of the surrounding soil (d) as a way to obtain nutrients. The amount of organic carbon deposited by root exudates varies by different plant species.

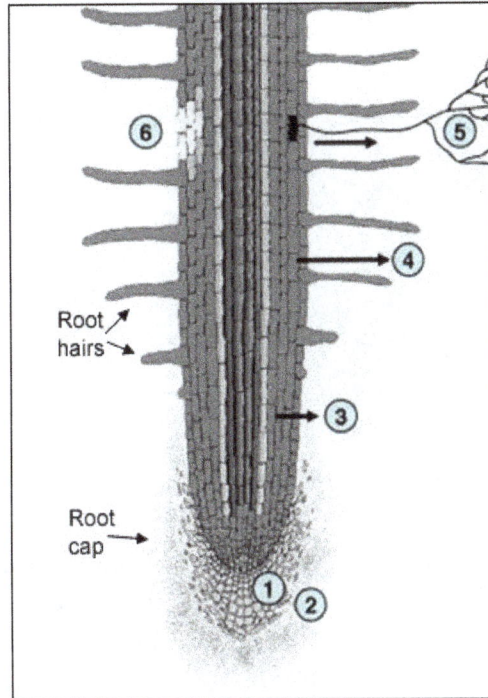

Diagram of a root in the rhizosphere showing six areas of rhizodeposition.

- Sloughing-off of outer root layers: The outer layer of roots sloughs-off as they push through the soil to reach down for nutrients and water. The sloughed-off layer of the root cap is released to the surrounding soil and becomes available for microbial decomposition.

References

- Permaculture: usrepresented.com, Retrieved 23 March, 2019

- Hay, R., and R. Porter. 2006 Physiology of Crop Yield (Second edition). Wiley-Blackwell. ISBN 1-4051-0859-2, ISBN 978-1-4051-0859-1

- Permaculture-importance-landscape-design: permaculturenews.org, Retrieved 28 June, 2019

- The-rhizosphere: permaculturenews.org, Retrieved 17 February, 2019

Desert Greening

8

- **Methods of Desert Greening**
- **Farmer-managed Natural Regeneration**
- **Afforestation**
- **Reforestation**

Man-made restoration of deserts for the purposes of farming, forestry or reclamation of natural water systems to support life, is defined as desert greening. It deals with methods such as, farmer-managed natural regeneration, afforestation, reforestation, etc. The topics elaborated in this chapter will help in gaining a better perspective about desert greening.

Desert greening is a concept where a number of methods are used to revitalize arid and semi-arid deserts. New methods are formulated every year. Here's a list of some of the revitalizing methods:

- Landscaping to reduce temperature, erosion, sandstorms, and evaporation.
- Greenhouse agriculture.
- Regeneration of salty, polluted, or degenerated soils.
- Permaculture - growing of plant communities.
- Planting trees and salt-loving plants such as Salicornia and Halophytes.
- Flood control.
- Seawater farming.
- Inland mariculture.
- Prevention of extensive grazing and firewood use.
- Training of local communities to care for plants, water, etc.

For example, the once lush Horqin Pasture in Mongolia has become a desert over time due to extensive grazing and deforestation brought about by a steady increase in the population of man and livestock. Lack of proper information on environment coupled with poor government support has led to the rapid deterioration of Horqin. As pastures disappear, farmers tend to move to other greenery, thus causing the Horqin Desert to expand at a rate of 10,000 km2 per year.

The U.S.-based global outdoor wear company, Timberland has taken up revitalizing of the Horqin Desert. As of April 2010, the company has helped plant 1 million trees and still counting. Their efforts are supported by Green Net. Together, they hope that tree planting aid in decreasing sandstorms in China and reversing mass desertification.

Droughts and deserts are a large part of Africa, and people tend to think that global warming will further aggravate the situation in the continent. However, things are contrary to what it seems. New evidence reveals that the rising temperatures are actually aiding Africans living around the driest parts of the continent.

Scientists are noticing that the Sahara desert and neighbouring areas are becoming greener due to the rise in rainfall. Satellite images of these regions show greenery developing in places like the Sahel, which is a semi-desert close to the Sahara with an area of 3860 km. According to a study in a journal , even central Chad and western Sudan are benefiting from this climate change. Experts believe that this is happening because hotter air can hold more moisture, thereby aiding to create more rain.

What has yet to be seen is sustainability. If the current trend continues, drought-ravaged regions could be revitalized and turned back to farmers. Well- established climate models reveal that this desert-shrinking trend could likely help in transforming Sahara into a lush savanna like it was 12,000 years ago.

Climate change is making man's basic necessities scarce in many parts of the world, thereby making water and food a very valuable commodity. The change is evident in the frequency and intensity of climatic turbulences we are witnessing every year.

Climate change can be combated as proved by Australian company, Greening the Desert™. They have been researching and dealing with drought, famine, salinity, and global warming issues quite successfully. Their recommended solution is to grow drought-resistant and salinity-tolerant cash crops, and to use saline waste water to irrigate the desert and drought regions such that carbon abatement is achieved in the process, which will help prevent global warming.

Methods of Desertgreening

Desert greening has the following methods:

- Managed intensive rotational grazing.

- Holistic management.

- Landscaping methods to reduce evaporation, erosion, consolidation of topsoil, sandstorms, temperature and more.

- Permaculture in general - harvesting runoff rainwater to grow plant communities polyculture, composting or multitrophic agriculture.

- Planting trees (pioneer species) and salt-loving plants such as Salicornia and Halophytes.

- Regeneration of salty, polluted, or degenerated soils.

- Floodwater retention and infiltration (flood control).

- Greenhouse agriculture.

- Seawater farming like done by the Seawater Foundation.

- Inland mariculture.

- Prevention of overgrazing and firewood use.

- Training of local residents to care for plantings, water systems etc.

- Planting trees with dew and rain harvesting technology like the Groasis Waterboxx.

- Farmer-managed natural regeneration.

Farmer-managed Natural Regeneration

Farmer-managed natural regeneration (FMNR) is a low-cost, sustainable land restoration technique used to combat poverty and hunger amongst poor subsistence farmers in developing countries by increasing food and timber production, and resilience to climate extremes. It involves the systematic regeneration and management of trees and shrubs from tree stumps, roots and seeds.

FMNR is especially applicable, but not restricted to, the dryland tropics. As well as returning degraded croplands and grazing lands to productivity, it can be used to restore degraded forests, thereby reversing biodiversity loss and reducing vulnerability to climate change. FMNR can also play an important role in maintaining not-yet-degraded landscapes in a productive state, especially when combined with other sustainable land management practices such as conservation agriculture on cropland and holistic management on range lands.

FMNR adapts centuries-old methods of woodland management, called coppicing and pollarding, to produce continuous tree-growth for fuel, building materials, food and fodder without the need for frequent and costly replanting. On farmland, selected trees are trimmed and pruned to maximise growth while promoting optimal growing conditions for annual crops (such as access to water and sunlight). When FMNR trees are integrated into crops and grazing pastures there is an increase in crop yields, soil fertility and organic matter, soil moisture and leaf fodder. There is also a decrease in wind and heat damage, and soil erosion.

In the Sahel region of Africa, FMNR has become a tool in increasing food security, resilience and climate change adaptation in poor, subsistence farming communities where much of sub-Saharan Africa's poverty exists. FMNR is also being promoted in East Timor, Indonesia, and Myanmar.

FMNR complements the evergreen agriculture, conservation agriculture and agroforestry movements. It is considered a good entry point for resource-poor and risk-averse farmers to adopt a low-cost and low-risk technique. This in turn has acted as a stepping stone to greater agricultural intensification as farmers become more receptive to new ideas.

Throughout the developing world, immense tracts of farmland, grazing lands and forests have become degraded to the point they are no longer productive. Deforestation continues at an alarming rate. In Africa's drier regions percent of rangelands and 61 percent of rain-fed croplands are damaged by moderate to very severe desertification. In some African countries deforestation rates exceed planting rates by 300.

Degraded land has an extremely detrimental effect on the lives of subsistence farmers who depend on it for their food and livelihoods. Subsistence farmers often make up to 70-80 percent of the population in these regions and they regularly suffer from hunger, malnutrition and even famine as a consequence.

In the Sahel region of Africa, a band of savanna which runs across the continent immediately south of the Sahara Desert, large tracts of once-productive farmland are turning to desert. In tropical regions across the world, where rich soils and good rainfall would normally assure bountiful harvests and fat livestock, some environments have become so degraded they are no longer productive.

Severe famines across the African Sahel in the 1970s and 1980s led to a global response, and stopping desertification became a top priority. Conventional methods of raising exotic and indigenous tree species in nurseries were used. Despite investing millions of dollars and thousands of hours of labour, there was little overall impact. Conventional approaches to reforestation in such harsh environments faced insurmountable problems and were costly and labour-intensive. Once planted out, drought, sand storms, pests, competition from weeds and destruction by people and animals negated efforts. Low levels of community ownership were another inhibiting factor.

Existing indigenous vegetation was generally dismissed as 'useless bush', and it was often cleared to make way for exotic species. Exotics were planted in fields containing living and sprouting stumps of indigenous vegetation, the presence of which was barely acknowledged, let alone seen as important.

This was an enormous oversight. In fact, these living tree stumps are so numerous they constitute a vast 'underground forest' just waiting for some care to grow and provide multiple benefits at little or no cost. Each stump can produce between 10 and 30 stems each. During the process of traditional land preparation, farmers saw the stems as weeds and slashed and burnt them before sowing their food crops. The net result was a barren landscape for much of the year with few mature trees remaining. To the casual observer, the land was turning to desert. Most concluded that there were no trees present and that the only way to reverse the problem was through tree planting.

Meanwhile, established indigenous trees continued to disappear at an alarming rate. In Niger, from the 1930s until 1993, forestry laws took tree ownership and responsibility for the care of trees out of the hands of the people. Reforestation through conventional tree planting seemed to be the only way to address desertification at the time.

Key Principles

FMNR depends on the existence of living tree stumps or roots in crop fields, grazing pastures, woodlands or forests. Each season bushy growth will sprout from the stumps/roots often appearing like small shrubs. Continuous grazing by livestock, regular burning and regular harvesting for fuel wood results in these 'shrubs' never attaining tree stature. On farmland, standard practice has been for farmers to slash this regrowth in preparation for planting crops, but with a little attention this growth can be turned into a valuable resource without jeopardising crop yields.

For each stump, a decision is made as to how many stems will be chosen to grow. The tallest and straightest stems are selected and the remaining stems culled. Best results are obtained when the farmer returns regularly to prune any unwanted new stems and side branches as they appear. Farmers can then grow other crops between and around the trees. When farmers want wood they can cut the stem(s) they want and leave the rest to continue growing. The remaining stems will increase in size and value each year, and will continue to protect the environment. Each time a stem is harvested, a younger stem is selected to replace it.

Various naturally occurring tree species can be used which may also provide berries, fruits and nuts or have medicinal qualities. In Niger, commonly used species include: Strychnos spinosa, Balanites aegyptiaca, Boscia senegalensis, Ziziphus spp., Annona senegalensis, Poupartia birrea and Faidherbia albida. However, the most important determinants are whatever species are locally available, their ability to re-sprout after cutting, and the value local people place on those species.

Faidherbia albida, also known as the 'fertiliser tree', is popular for intercropping across the Sahel as it fixes nitrogen into the soil, provides fodder for livestock, and shade for crops and livestock. By shedding its leaves in the wet season, Faidherbia provides beneficial light shade to crops when high temperatures would otherwise damage crops or retard growth. Leaf fall contributes useful nutrients and organic matter to the soil.

The practice of FMNR is not confined to croplands. It is being practised on grazing land and in degraded communal forests as well. When there are no living stumps, seeds of naturally occurring species are used. In reality, there is no fixed way of practising FMNR and farmers are free to choose which species they will leave, the density of trees they prefer, and the timing and method of pruning.

In Practice

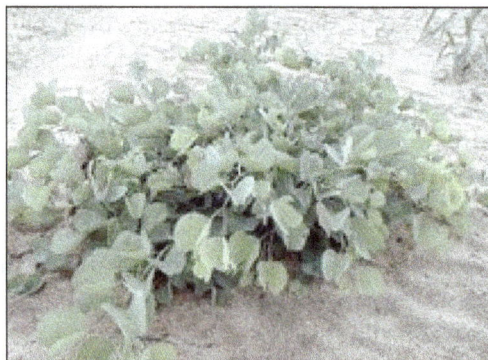

FMNR depends on the existence of living tree stumps, tree roots and seeds to be re-vegetated. These can be in crop fields, grazing lands or degraded forests. New stems, which sprout from these stumps and tree roots, can be selected and pruned for improved growth. Sprouting tree stumps and roots may look like shrubs and are often ignored or even slashed by farmers or foresters.

However, with culling of excess stems and by selecting and pruning of the best stems, the regrowth has enormous potential to rapidly grow into trees.

Seemingly treeless fields may contain seeds and living tree stumps and roots which have the ability to sprout new stems and regenerate trees. Even this 'bare' millet field in West Africa contains hundreds of living stumps per hectare which are buried beneath the surface like an underground forest.

Step 1: Do not automatically slash all tree growth, but survey your farm noting how many and what species of trees are present.

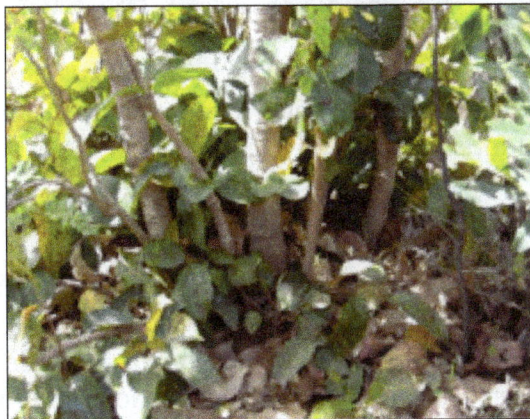

Step 2: Select the stumps which will be used for regeneration.

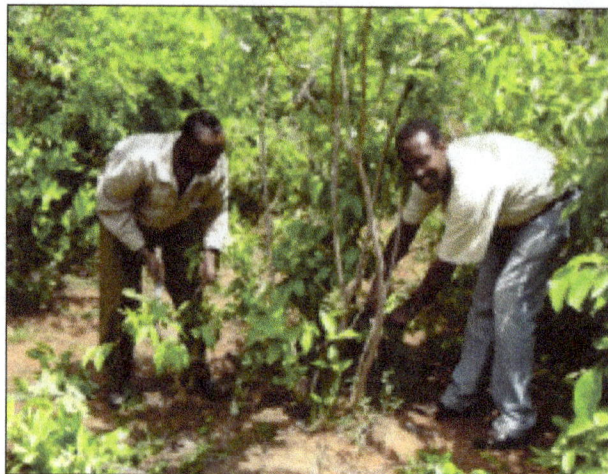

Step 3: Select the best five or so stems and cull unwanted ones. This way, when you want wood you can cut the stems that are needed and leave the rest to continue growing. These remaining stems will increase in size and value each year, and will continue to protect the environment and provide other useful materials and services such as fodder, humus, habitat for useful pest predators and protection from the wind and sun. Each time one stem is harvested, a younger stem is selected to replace it.

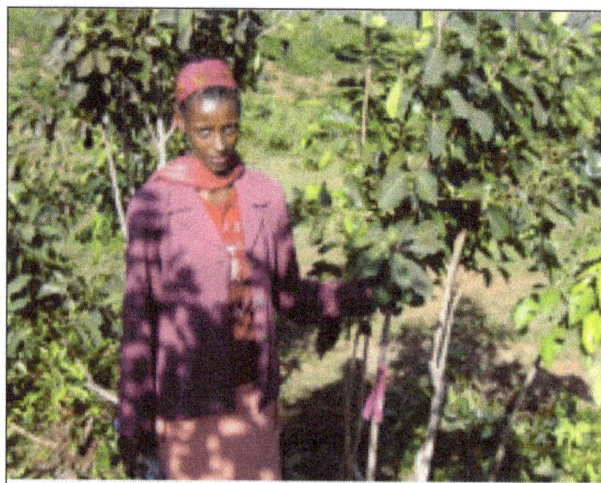

Tag selected stems with a coloured rag or paint. Work with the whole community to draw up and agree on laws which will protect the trees being pruned and respect each person's rights. Where possible, include government forestry staff and local authorities in planning and decision making.

Benefits

FMNR can restore degraded farmlands, pastures and forests by increasing the quantity and value of woody vegetation, by increasing biodiversity and by improving soil structure and fertility through leaf litter and nutrient cycling. The reforestation also retards wind and water erosion; it creates windbreaks which decrease soil moisture evaporation, and protects crops and livestock against searing winds and temperatures. Often, dried up springs reappear and the water table rises towards historic levels; insect eating predators including insects, spiders and birds return, helping to keep crop pests in check; the trees can be a source of edible berries and nuts; and over time the biodiversity of plant and animal life is increased. FMNR can be used to combat deforestation and desertification and can also be an important tool in maintaining the integrity and productivity of land that is not yet degraded.

Trials, long-running programs and anecdotal data indicate that FMNR can at least double crop yields on low fertility soils. In the Sahel, high numbers of livestock and an eight month dry season can mean that pastures are completely depleted before the rains commence. However, with the presence of trees, grazing animals can make it through the dry season by feeding on tree leaves and seed pods of some species, at a time when no other fodder is available. In northeast Ghana, more grass became available with the introduction of FMNR because communities worked together to prevent bush fires from destroying their trees.

Well designed and executed FMNR projects can act as catalysts to empower communities as they negotiate land ownership or user rights for the trees in their care. This assists with self-organisation, and with the development of new agriculture-based micro-enterprises (e.g., selling firewood, timber and handcrafts made from timber or woven grasses).

Conventional approaches to reversing desertification, such as funding tree planting, rarely spread beyond the project boundary once external funding is withdrawn. By comparison, FMNR is cheap, rapid, locally led and implemented. It uses local skills and resources – the poorest farmers can learn by observation and teach their neighbours. Given an enabling environment, or at least the absence of a 'disabling' environment, FMNR can be done at scale and spread well beyond the original target area without ongoing government or NGO intervention.

World Vision evaluations of FMNR conducted in Senegal and Ghana in 2011 and 2012 found that households practising FMNR were less vulnerable to extreme weather shocks such as drought and damaging rain and wind storms.

The following table summarises FMNR's benefits which fit the sustainable development model of economic, social and environmental benefits:

Economic benefits	Social benefits	Environmental benefits
Increased crop yields (often double or triple).	Increased food security and nutrition (including native fruits, nuts and seeds).	Reduced erosion.
Increased fodder from edible leaves and seed pods, and in some cases increased pasture growth.	Less distance for women and children to travel to collect firewood.	Reduced soil-moisture evaporation due to wind breaks shading and mulching.
Higher livestock productivity and survival.	Community capacity building to deal with local, regional and national governments and regulators.	Increased soil fertility.
Reduced impact from floods and drought – trees provide alternative income and livelihood sources making impacts less severe and recovery faster.	Improved governance through clarification of tree ownership laws and regulations.	Improved soil structure through greater quantities of organic matter.
Increased income generation through diversification (e.g. timber and fuel wood) and intensification of activities.	Education and training in farming and marketing.	Increased water infiltration and groundwater recharge.

Economic flow-on effects such as employment and greater purchasing capacity.	Reduced need for migration by young people and men to cities.	Increased biodiversity, environmental restoration and tree cover.
Increased economic activity creates opportunities, e.g., development of new business models such as cooperatives.	Higher incomes result in better opportunities for medical treatment, children's education, nutrition and clothing, etc.	Enhanced resilience to climate change.
	Empowerment for community members to live independently with hope for the future.	

Key Success Factors and Constraints

While there are numerous accounts of the uptake and spread of FMNR independent of aid and development agencies, the following factors have been found to be beneficial for its introduction and spread:

- Awareness creation of FMNR's potential.

- Capacity building through workshops and exchange visits.

- Awareness of the devastating effects of deforestation. The adoption of FMNR is more likely when communities acknowledge their situation and the need to take action. This perception of need can be supported by education.

- An FMNR champion/facilitator from within the community who encourages, challenges and trains peers. This is critical during the first three to five years, and continues to be important for up to 10 years. Regular site visits also ensure early detection and remedial action on resistance and threats to FMNR through deliberate damage to trees and theft.

- The buy-in of all stakeholders including their agreement on any by-laws created for FMNR and the consequences for infringements. Stakeholders include FMNR practitioners, local, regional and national government departments of agriculture and forestry, men, women, youth, marginalised groups (including nomadic herders), cultivators and commercial interests.

- Stakeholder buy-in is also important to create a critical mass of FMNR adopters in order to change social attitudes from a position of apathy or active participation in deforestation to one of proactive sustainable tree management through FMNR.

- Government support through the creation of favourable policies, positive reinforcement of actions facilitating the spread of FMNR, and disincentives for actions working against the spread of FMNR. FMNR practitioners need to be confident that they will benefit from their labours (either private or community ownership of trees, or legally binding user rights).

- Reinforcement of existing organisational structures (farmers clubs, development groups, traditional leadership structures) or establishment of new structures which will provide a framework for communities to practise FMNR on a local, district or region-wide basis.

- A communications strategy which includes education in schools, radio programs and engagement with religious and traditional leaders to become advocates.

- Establishment of a legal, transparent and accessible market for FMNR wood and non-timber forest products, enabling practitioners to benefit financially from their activities.

The two main reasons why FMNR has spread so widely in Niger are attitudinal change by the community of what constitutes good land management practices, and farmers' ownership of trees. Farmers need the assurance that they will benefit from their labour. Giving farmers either outright ownership of the trees they protect, or tree-user rights, has made it possible for large-scale farmer-led reforestation to take place.

Current and Future Directions

Over nearly 30 years, FMNR has changed the farming landscape in some of the poorest countries in the world, including parts of Niger, Burkina Faso, Mali, and Senegal, providing subsistence farmers with the methods necessary to become more food secure and resilient against severe weather events.

The 2011–2012 food crisis in East Africa gave a stark reminder of the importance of addressing root causes of hunger. In the 2011 State of the World Report, Bunch concludes that four major factors – lack of sustainable fertile land, loss of traditional fallowing, cost of fertiliser and climate change – are coming together all at once in a sort of "perfect storm" that will almost surely result in an African famine of unprecedented proportions, probably within the next four to five years. It will most heavily affect the lowland, semi-arid to sub-humid areas of Africa (including the Sahel, parts of eastern Africa, plus a band from Malawi across to Angola and Namibia); and unless the world does something dramatic to 30 million people could die from famine between 2015 and 2020. Restoration of degraded land through FMNR is one way of addressing these major contributors to hunger.

In recent years FMNR has come to the attention of global development agencies and grassroots movements alike. The World Bank, World Resources Institute, World Agroforestry Center, USAID and the Permaculture movement are amongst those either actively promoting or advocating for the uptake of FMNR and FMNR has received recognition from a number of quarters including:

- In 2010, FMNR won the Interaction 4 Best Practice and Innovation Initiative award in recognition of high technical standards and effectiveness in addressing the food security and livelihood needs of small producers in the areas of natural resource management and agro forestry.

- In 2011, FMNR won the World Vision International Global Resilience Award for the most innovative initiative in the area of resilient development practice and natural environment and climate issues.

- In 2012, WVA was awarded the Arbor Day Award for Education Innovation.

In April 2012, World Vision Australia – in partnership with the World Agroforestry Center and World Vision East Africa – held an international conference in Nairobi called "Beating Famine" to analyse and plan how to improve food security for the world's poor through the use of FMNR and Evergreen Agriculture. The conference was attended by more than 200 participants, including world leaders in sustainable agriculture, five East African ministers of agriculture and the environment, ambassadors, and other government representatives from Africa, Europe, and Australia, and leaders from non-government and international organisations.

Two major outcomes of the conference were:

- The establishment of a global FMNR network of key stakeholders to promote, encourage and initiate the scale-up of FMNR globally.

- Country, regional and global level plans as a basis for inter-organisation collaboration for FMNR scale-up.

The conference acted as a catalyst for media coverage of FMNR in some of the world's leading outlets and a noticeable increase in momentum for an FMNR global movement. This heightened awareness of FMNR has created an opportunity for it to spread exponentially worldwide.

Afforestation

Afforestation is the process of planting trees, or sowing seeds, in a barren land devoid of any trees to create a forest. The term should not be confused with reforestation, which is the process of specifically planting native trees into a forest that has decreasing numbers of trees. While reforestation is increasing the number of trees of an existing forest, afforestation is the creation of a 'new' forest.

Our Earth has been constantly trying to cope with the way in which human beings use natural resources, clear forest lands, cut trees, and contaminate the air, land, and water. Industrial revolution, population bursts, and pollution create permanent damage to the earth, and the result is global warming and climate change. In such situations, something that can help extend the life of the planet and its living organisms is the increase of natural resources and decrease of exploitation of these resources.

By planting trees and creating forests, many of the commercial needs of human beings are fulfilled, while not destroying what is left of the planet. Afforestation is, therefore, a practice that has been propagated by government and non-government agencies of many countries as a way to stop over-exploitation of nature.

Importance

The importance is immense in today's scenario because it is mainly done for commercial purposes. In a natural forest or woodland, the trees are heterogeneous. Owing to the sensitivity to over usage and slow growths, these forests cannot be used continuously for commercial purposes like wood products. The process of planting trees in empty lands helps promote the fast propagation of specific types of trees for the wood industry.

With the increasing demand for wood fuels and building materials, this process helps to meet these demands without cutting down the natural forests. Deforestation can lead to the depletion of trees in water catchments and riverside zones. Afforestation ensures trees and plants that hold the soil in these sensitive areas remain protected.

Many countries have introduced the practice of planting trees along with agricultural crops in croplands. The benefits of this practice, which is called agroforestry, are:

In terms of the environmental benefits, planting trees is always beneficial whether it takes place in a barren land or is used as a method to regenerate a depleted forest. Trees help check atmospheric carbon dioxide; large scale afforestation can curb the problems caused due to burning of fossil fuels, industrialization and so forth.

Current Efforts

In the central hardwood forest region of the US, increasing numbers of land owners are converting crop land marginally into a forest. This is being done to decrease the pressure on the use of existing hardwood species of the forest like black cherry, black walnut, and northern red oak.

In South Africa, about 0.5 percent of land is covered with indigenous forests, and 1.1 percent by forests formed by Total Commercial Afforestation (TCA) and containing trees like pine, gum trees, black wattle, and so forth. This has helped provide wood to be used for charcoal, poles, mining timber, paper pulp, and other commercial applications.

The advantage of planting a tree species, like pine, is it helps check infections the tree is prone to in its native country and climate, thus producing higher production. Pursuant to better growth and higher yields due to afforesting of these alien species, South Africa can produce and export close to two million tons of wood and wood products.

In China, the government has earmarked a bulk amount equivalent to almost 300 billion US dollars that would be completely utilized for afforesting schemes the country is planning. To combat soil erosion in Central and West China, the government has already started the process of converting farmland back to woodland.

Reforestation

Reforestation is the process of regenerating or replanting forest areas that have been destroyed or damaged for the benefits of mankind. Reforestation and afforestation share the same meaning i.e. afforestation is another name given to reforestation. Occasionally forests have the capability to regenerate due to the trees in the surroundings or due to the dispersion of seeds. However, forest lands that are badly degraded cannot be regenerated unless plants have been planted by using native methods.

Why Reforestation?

Reforestation is a very important procedure in order to save our planet. This is needed as huge forests are being destroyed or damaged due to various reasons on a daily basis. Removal of the green cover from the surface of the earth has become common due to various reasons such as forest fires, agricultural needs, human needs, logging, and mining.

Deforestation

Forests play a very important role in order to maintain the balance in the cosystem. Deforestation is a serious threat to the existence of mankind. Due to deforestation, serious issues have risen like the greenhouse effectdue to excessive carbon compounds present in the air. Forests house a diverse range of plants and animals. They provide a livelihood for a huge number of people. Forests also play a vital role in maintaining the water cycle, preventing soil erosion. Most important of all, forests are responsible for maintaining the balance of carbon dioxide and oxygen the in earth's atmosphere. Therefore, reforestation plays an important role in order to overcome deforestation and to restore the natural balance of plant life on the planet.

Large areas of forests have been disturbed. The Atlantic forest situated in South America is the house for a wide variety of wildlife and includes 104 species which are found nowhere else on earth. This forest is in great danger and can be considered as one of the most disturbed ecosystems in the world. This forest is severely fragmented and fails to support a large number of species. Thus justifying the importance of trees to the planet.

How are we Implementing Reforestation?

Many organizations are working towards the protection and restoration of forests by various methods such as imparting education to people regarding the importance of forests, reforestation, and expansion of the protected areas. Governments in different countries are also trying to introduce strict policies regarding the protection and restoration of forests. A collective global effort is required in order to achieve a sustainable and balanced ecosystem.

PERMISSIONS

We would like to thank the editorial team for lending their expertise to make the book truly unique. They have played a crucial role in the development of this book. Without their invaluable contributions this book wouldn't have been possible. They have made vital efforts to compile up to date information on the varied aspects of this subject to make this book a valuable addition to the collection of many professionals and students.

This book was conceptualized with the vision of imparting up-to-date and integrated information in this field. To ensure the same, a matchless editorial board was set up. Every individual on the board went through rigorous rounds of assessment to prove their worth. After which they invested a large part of their time researching and compiling the most relevant data for our readers.

The editorial board has been involved in producing this book since its inception. They have spent rigorous hours researching and exploring the diverse topics which have resulted in the successful publishing of this book. They have passed on their knowledge of decades through this book. To expedite this challenging task, the publisher supported the team at every step. A small team of assistant editors was also appointed to further simplify the editing procedure and attain best results for the readers.

Apart from the editorial board, the designing team has also invested a significant amount of their time in understanding the subject and creating the most relevant covers. They scrutinized every image to scout for the most suitable representation of the subject and create an appropriate cover for the book.

The publishing team has been an ardent support to the editorial, designing and production team. Their endless efforts to recruit the best for this project, has resulted in the accomplishment of this book. They are a veteran in the field of academics and their pool of knowledge is as vast as their experience in printing. Their expertise and guidance has proved useful at every step. Their uncompromising quality standards have made this book an exceptional effort. Their encouragement from time to time has been an inspiration for everyone.

The publisher and the editorial board hope that this book will prove to be a valuable piece of knowledge for students, practitioners and scholars across the globe.

INDEX

www.ingramcontent.com/pod-product-compliance
Lightning Source LLC
Chambersburg PA
CBHW082023190326

41458CB00010B/3251